UNIFIX® TEACHER'S RESOURCE BOOK

LOLA J. MAY
LARRY ECKLUND

Didax
Educational Resources

UNIFIX

Contents

Introduction

From Concrete to Abstract with Manipulatives

The Curriculum and Evaluation Standards for School Mathematics, a publication of the National Council of Teachers of Mathematics, emphasizes that the mathematics curriculum should involve students in doing mathematics.

Children are active individuals who construct, modify and integrate ideas by interacting with the physical world, real materials and other children. With this in mind, it is clear that learning mathematics must be an active process. In the Curriculum and Evaluation Standards, verbs such as explore, construct, use, investigate, justify, represent, solve, discuss, describe, develop and predict are used to convey this active physical and mental involvement of children in learning mathematics.

Although generalizations and abstractions are an integral part of mathematics learning, research has shown that there is no one best way to teach them. There are, however, several learning theories, and each of these theories recognizes the importance of the concrete level of learning that uses manipulatives and hands-on activities to lead to generalization and abstraction.

We know that learning occurs from both physical and mental activity and that abstraction in mathematics is a slow process requiring years of development. We also know that many individual differences exist and that learning routes and the rate of progress differ from individual to individual and varies greatly among groups of students.

The essential role of teachers is to help learners build bridges from the concrete to the abstract. In performing this role, the teacher must make effective use of appropriate models that involve students in experiencing, learning and eventually abstracting mathematics.

Unifix Materials

Brightly colored and structurally very durable, interlocking Unifix Cubes are ideal for visually appealing, concrete presentations of mathematical concepts. The cubes and the Unifix ancillary materials are designed for pre-number and early number work as well as more advanced mathematics activities.

Unifix Cubes are unit based and can be used individually or locked into structural bars of different lengths enabling students to discover, experiment and learn a wide variety of mathematical concepts. The cubes can be used in color classification activities, one-to-one correspondence, patterning, sets and counting, ordering and operations.

The flexibility of Unifix materials allows them to be used with students from preschool to intermediate levels and with special education students.

Unifix Teacher's Resource Book

The Unifix Teacher's Resource Book has been written to help you use your Unifix materials in the most effective manner possible and to introduce you to the many ways Unifix materials can be used to teach mathematical concepts. The materials have a greater versatility than most teachers are aware of, and it is the intent of this resource book to present a variety of activities and to stimulate your own creativity in using Unifix materials.

The Unifix Teacher's Resource Book is arranged by level of difficulty with activities that give students in the primary grades and students in need of remedial work an opportunity to work with manipulatives and internalize the mathematical concepts being presented.

Each chapter contains lessons that provide several methods of approaching the subject area and the chapters progress in difficulty, beginning with basic Building Patterns and Graphs to Multiplication and Division.

The Index has been arranged so that you can find the chapters and lessons that use the specific Unifix materials that you may already have on hand. That is, if you have a supply of Unifix 100 Tracks, you can look in the Index to find the lessons that use that material.

You will also find photographs of all the Unifix materials used throughout this book on pages 64 and 65.

The learning process is really a never ending one. Although we are "old Unifix hands," working together on this resource book has helped us to see even more ways to demonstrate mathematical concepts with manipulatives in general and Unifix materials in particular. We hope you will enjoy using this book as much as we enjoyed writing it.

We wish to thank Arthur Wiebe and Sue Atkinson for their helpful advice. We are also grateful to Knud Borchersen of the publishers, Didax Educational Resources, Dee Corr, editor, and Alan Cassedy, designer, for their support in making the publication of this book possible.

Lola J. May

Larry Ecklund

UNIFIX

On Teaching Elementary Mathematics

UNIFIX

Addition

The main concept to be conveyed to young students regarding the operation of addition is that when adding, two or more sets are combined together to make a combined set. This concept is also described as "part plus part equals whole."

Manipulatives will help the student to understand this concept by feel and touch. For example, the young student will experience that three objects and one more object makes four objects every time the objects are manipulated. Unifix cubes are particularly helpful to illustrate this experience.

Familiarity with these early number facts is one of the main avenues by which a student will develop confidence with manipulatives.

The language of addition forms a critical part of the early learning of number facts. Young students should be encouraged to talk about what they are doing as they manipulate materials, such as one student may explain to another student what he or she is doing, *"Three cubes and four cubes are seven cubes altogether."*

3
+4
7

Using the language of addition helps students understand the written number sentence, 3 + 4 = 7.

The student will also become aware that he or she can add in any order (3 + 6) + 2 is the same as 3 + (6 + 2) as he or she manipulates cubes in many different situations. This is called the associative property of addition.

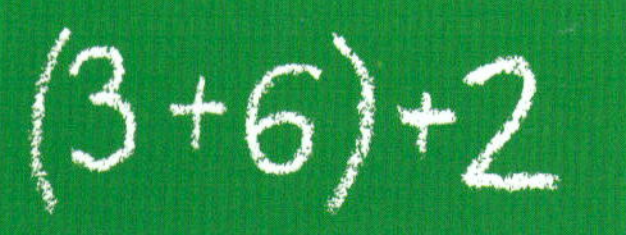

It is also important that students see numbers as a number line. As well as an obvious aid in the learning of counting on, the number line becomes an important part of the students' mental imagery of mathematics, helps develop an in-depth understanding of math and aids in mental calculations. In chapters 4, 5 and 8 you will find further notes and lesson plans illustrating these concepts.

Place Value

The concept of place value is critical to an understanding of number work. The three place value concepts that need special attention in the early stages of mathematics instruction are grouping, positioning and trading. No amount of written arithmetic on its own will give students adequate understanding of these place value concepts; they must be learned through manipulative activities and through talking about and explaining those activities.

Although we mainly group in tens, we must not forget that young students also need practice in equal groupings of other numbers.

However, the idea of grouping in tens must be emphasized when counting large numbers. This leads naturally to the use of money and other activities with many of the Unifix materials. Activities with Unifix materials also make the need to trade clear to the students.

Using the Hundreds, Tens and Ones Place Value Tray and the Unifix Cubes and Tens Cubes, we can demonstrate the concept of equivalence to the students when we trade ten cubes for one Ten Cube.

An easily understood analogy of the concept of equivalence that young students understand is that ten pennies equal one dime.

To demonstrate position, the Place Value Tray and the Hundreds and Ten Cubes can be used to show that a blue cube in the tens column is of higher value than a cube in the ones column and that a red cube in the hundreds column has a higher value than one in the tens column.Chapters 3 and 7 provides notes and lesson plans illustrating place value concepts.

Subtraction

Subtraction is the inverse operation of addition in which we split or partition a set of numbers. Using a manipulative such as Unifix Cubes, young students can see that five cubes can be split up into a set of four and a set of one or a set of three and a set of two, etc.

The language patterns used in subtraction are complex, but it is only as these are explored and discussed that the young student estab-ishes a true understanding of subtraction.

Some of the language used while working with subtraction manipulatives includes:

8 is 5 more than 3, 5 is 3 less than 8.

If I count back 3 from 8 on the number line, where will I be?

The difference between 8 and 3 is 5.

How much more than 5 is 8?

If I had $8.00 and spent $5.00, how much is left?

In chapters 6 and 9 you will find introductory notes and lesson plans on the meaning of subtraction.

Multiplication

The earliest stages of multiplication are repeated addition. It is this recognition and counting of equivalent sets that underlies the idea of multiplication as repeated addition.

The extension of this idea of multiple addition of equivalent sets can be demonstrated with the Unifix 100 Track and Markers or by making "trains" of cubes in patterns of colors such as three red, three blue, three red, three blue.

Later these four trains can be rearranged into a rectangular array illustrating four set of three.

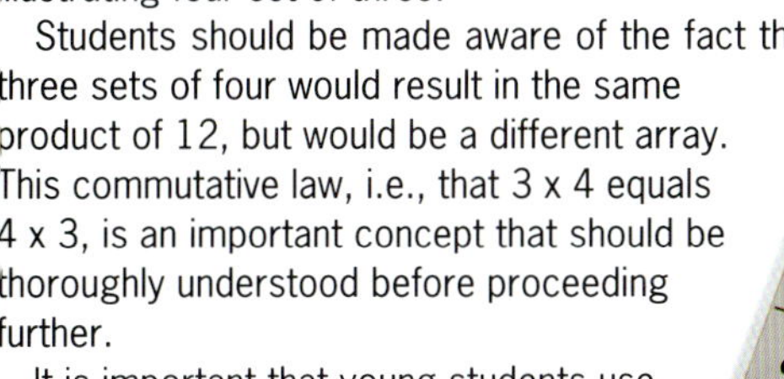

Students should be made aware of the fact that three sets of four would result in the same product of 12, but would be a different array. This commutative law, i.e., that 3 x 4 equals 4 x 3, is an important concept that should be thoroughly understood before proceeding further.

It is important that young students use the specific multiplication language while working with manipulatives so that the concepts will be fully understood. Students need time to explore the language and to verbalize what they are doing before recording the operation in any written form. It is the exploration and the use of the language at this early stage that lays the sound foundation for later learning.

Chapters 10, 11 and 12 offer additional introductory notes and several lessons on multiplication.

Division

Just as subtraction is the inverse operation of addition, division is the inverse operation of multiplication.

Division is a complex concept and is, therefore, introduced after multiplication. In presenting division, the idea of sharing things equally or of "fair share" is one with which young students can relate and should be the basis for beginning division work. This "sharing" aspect of division is the easiest for the student to grasp at first.

The other meaning of division, grouping, although easy to demonstrate using manipulatives can be confusing to students. This aspect of division, when the number in each subset is known, can best be explained using concrete examples.

In sharing, the number in the total set and the number of subsets is known. The unknown number is the size of each share. The example of distributing a number of cookies equally between a certain number of children can be used. For example, I have 12 cookies to share fairly among six children. How many cookies will each child receive?

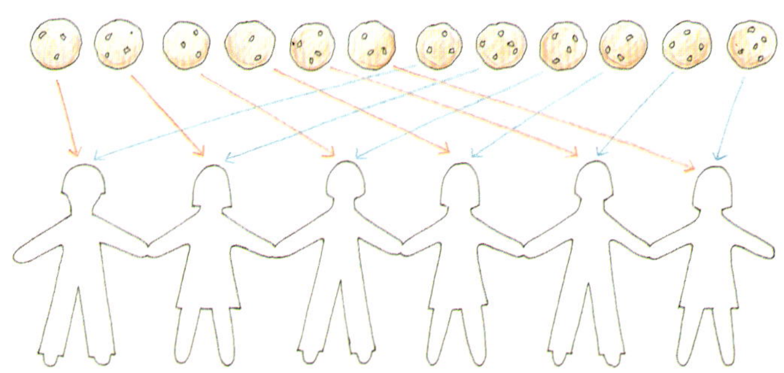

Each child will receive two cookies.

In grouping, the size of the total set and the size of the subset is known, but the number of sets is unknown. Using the example of distributing cookies, we want to give each child two cookies so we know the size of the group, but now we must determine how many children can have two cookies each. Grouping problems can be solved with repeated subtraction.

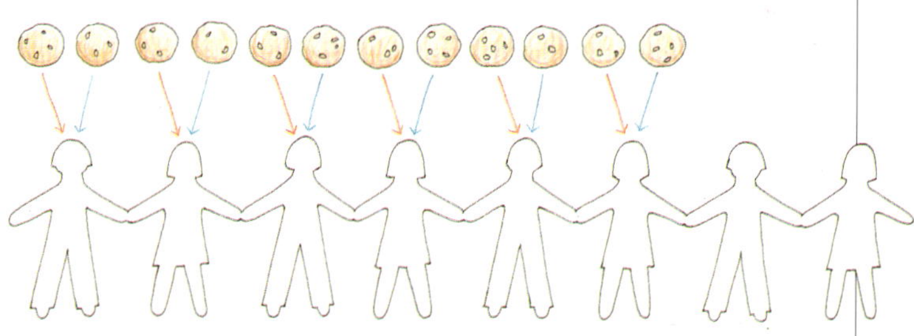

One of the problems with division for students is that until multiplication is fully understood, the student will usually be unable to work out division numerically. It is, therefore, important to use manipulative materials such as the 100 Track to illustrate repeated subtraction and to make trains of cubes. Just as five trains of three make 15, so 15 can be divided back into five trains of three.

Chapter 13 has a brief introduction and lesson notes offering further insights into the meaning of division.

UNIFIX

1 Building Patterns

The study of mathematics is made up of patterns, numbers and the properties of geometric shapes. As a beginning in their understanding of mathematics, students need to become aware of and create patterns.

Unifix Cubes are excellent manipulative materials that can encourage early creating of patterns. The cubes allow students to build patterns by interlocking the cubes together in linear color patterns (*red - blue, red - blue, red - blue*), into towers and other patterns and shapes.

Putting Unifix Cubes together and snapping them apart allows students to feel a sense of accomplishment as they explore pattern making on their own as well as constructing patterns using pattern underlay cards.

There are four essential steps in helping students learn about and understand patterns:

- Free exploration
- Recognizing and repeating a pattern
- Extending a pattern
- Completing a pattern

These four steps have been incorporated in the following six lesson plans.

Cubes

Cube Corner Units

Rod Stamps

Cubes for the Overhead Projector

Wax Crayons

Operational Grid and Tray

Pattern Building Underlay Cards

100 Track

Gummed Sheets

Building Patterns

Lesson 1

Objective:

During free exploration time, students will become familiar with Unifix Cubes, how they snap together and come apart and will be encouraged to create designs and patterns.

Materials:

UNIFIX CUBES

UNIFIX CUBE CORNER UNITS

Activities:

Provide each student with twenty *Unifix Cubes*, ten each of two colors, and a number of *Corner Units*, and let him or her explore the new media independently at his or her desk or with two or three students at the learning center.

Encourage them to play with the material.

Students need many experiences of exploring how the cubes fit together and how to make different patterns and designs. They will want to tell their peers about how they put their designs together and will want to copy what their friends have made.

It is important that you move around the room from group to group and listen to the students as they manipulate the Unifix Cubes.

It is also important to provide time for the students to share with each other the designs and patterns they have created with the cubes.

Lesson 2

Objective:

Students will recognize and repeat patterns demonstrated by the teacher.

Materials:

UNIFIX CUBES

UNIFIX CUBES FOR THE OVERHEAD PROJECTOR

UNIFIX 100 TRACKS

PATTERN BUILDING UNDERLAY CARDS

OPERATIONAL GRIDS AND TRAYS

Activities:

Distribute a quantity of *Unifix Cubes, Pattern Building Underlay Cards and Operational Grids and Trays* to the students.

Make a simple linear pattern using Unifix Cubes. Repeat the pattern.

Display the pattern on the chalkboard ledge, on the 100 Track or at the learning center table. If an overhead projector is available, you can also use the *Unifix Cubes for the Overhead Projector* to demonstrate the pattern.

It is important that the students recognize the pattern and are able to see that there is a repeat of the placement of the cubes.

Ask the students which color cube is at the beginning and ending of the pattern. The students can demonstrate the pattern by clapping their hands and patting their laps as you point to each cube, e.g., for a pattern of *blue, blue, red, red, red,* the students would *clap, clap, pat, pat, pat.*

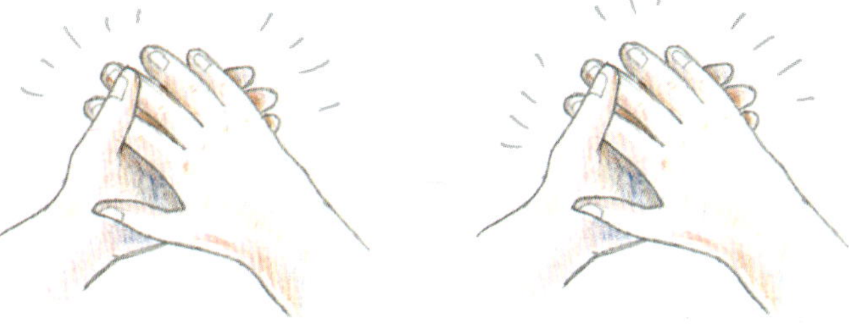

Then encourage the students to copy the pattern with their cubes and then to create an original pattern of their own.

Ask them to repeat their pattern.

As a small group activity, ask the students to work together to complete the patterns using the *Pattern Building Underlay Cards* and the *Operational Grid and Tray*.

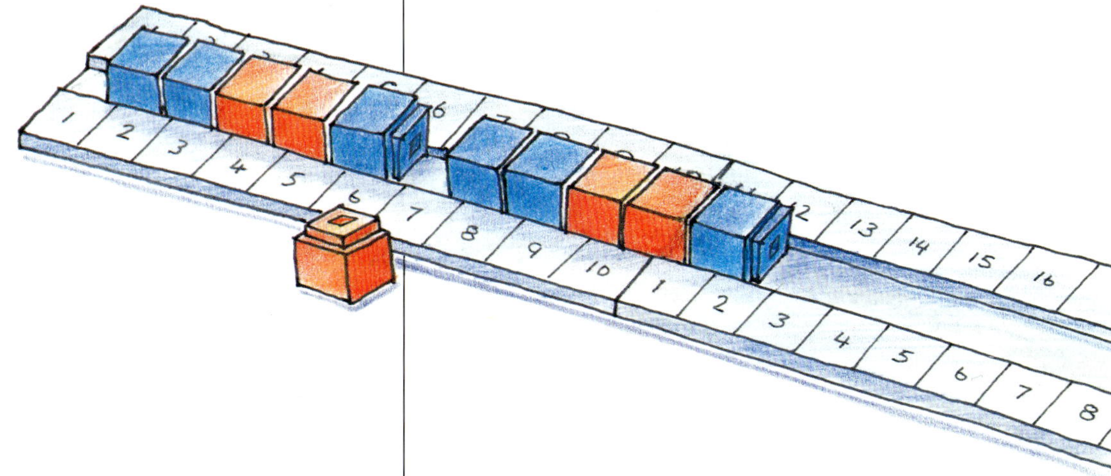

Lesson 3

Objective:

Students will copy a pattern a third time after the teacher has demonstrated a pattern and repeated it.

Materials:

UNIFIX CUBES
UNIFIX 100 TRACKS
UNIFIX BLANK UNDERLAY CARDS
UNIFIX CUBES FOR THE OVERHEAD PROJECTOR
UNIFIX WAX CRAYONS

Activities:

Distribute a quantity of *Unifix Cubes* and *100 Tracks* or *Blank Underlay Cards* and *Wax Crayons.*

Build a pattern with the Unifix Cubes or show a pattern by using the *Cubes for the Overhead Projector* on the overhead projector. Repeat the pattern.

Ask the students to copy the pattern the third time at their desks by using the cubes on a portion of the 100 Track or by copying and coloring the pattern on a Blank Underlay Card with Wax Crayons.

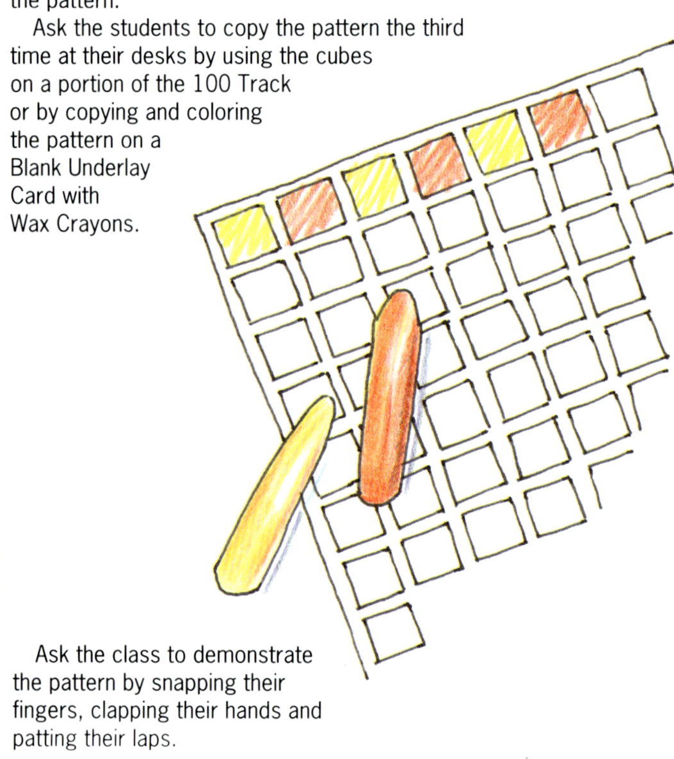

Ask the class to demonstrate the pattern by snapping their fingers, clapping their hands and patting their laps.

Students can next work in pairs. Each one can make a pattern, repeat it and then exchange the pattern with his or her partner who copies the pattern the third time.

Lesson 4

Objective:

Students will recognize the pattern in two sequences and complete the unfinished pattern in the third sequence.

Materials:

UNIFIX CUBES
UNIFIX 100 TRACKS
UNIFIX ROD STAMPS
UNIFIX WAX CRAYONS

Activities:

In this next stage, using *Unifix Cubes* make a pattern and repeat the pattern. Then start the pattern a third time and ask the students to tell which cube or cubes would be next in the pattern.

Distribute cubes and the *100 Track* to the students. Using the 100 Track, ask the students to work in pairs each making a pattern, repeating it and starting the pattern a third time and having his or her partner complete the pattern.

As an extended activity, using the *Unifix Rod Stamps* and the *Unifix Wax Crayons*, ask the students to make their own unfinished patterns for their partners to complete.

These patterns can be displayed on the bulletin board or assembled into a pattern book.

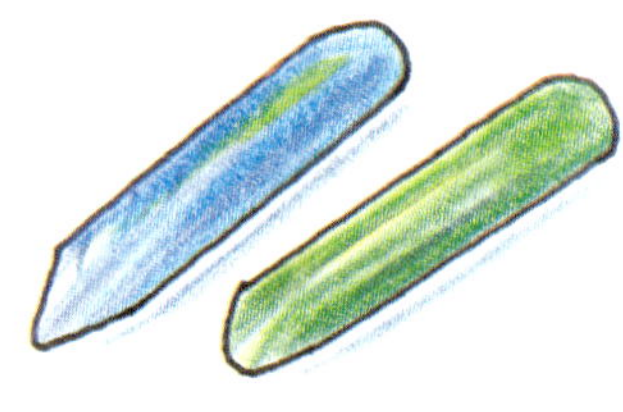

Building Patterns

Lesson 5

Objective:

Using the Pattern Building Underlay Cards, students will recognize patterns, see the symmetry of the pattern and copy a non-linear pattern.

Materials:

UNIFIX CUBES
UNIFIX OPERATIONAL GRIDS AND TRAYS
UNIFIX PATTERN BUILDING UNDERLAY CARDS
UNIFIX BLANK UNDERLAY CARDS
UNIFIX WAX CRAYONS

Activities:

This activity can be conducted with small groups of students in which all students should be encouraged to participate.

To introduce the pattern cards, display several cards in which there are obvious patterns such as a cross, checkerboard, windows, etc.

Ask the students to point out what they see in each pattern card and describe the pattern.

What happens to the pattern if the card is turned on its side? If turned upside down?

Select a *Pattern Building Underlay Card* in which the pattern is identical at the top and bottom, that is, the pattern would be replicated if folded over on itself.

Place the Pattern Building Underlay Card in the *Operational Tray* and place the Grid over the card. Cover the bottom half of the grid and ask the students to cooperatively complete the top half of the pattern using *Unifix Cubes.*

After the pattern is completed, lift the grid out of the tray and place the grid with cubes on the lower half of the pattern card. Ask the students to tell what has happened. Be sure students are aware that the pattern at the top of the card is the same as the pattern at the bottom of the card.

Give students a supply of Unifix Cubes and ask each student to make a copy of the pattern in their Operational Grids and Trays. It may help to have a student point to each cube in a row and say its *"name"* such as *"blue, blue, yellow, yellow, etc."* After the patterns have been completed, ask the students to compare their patterns with the pattern on the underlay card.

As a further extension of this activity, students can use a Blank Underlay Card and crayons to make a permanent copy of the pattern.

Lesson 6

Objective:

Using the Pattern Building Underlay Cards, students will recognize and duplicate patterns while gaining an understanding of conservation of number.

Materials:

UNIFIX CUBES
UNIFIX OPERATION GRID AND TRAY
UNIFIX PATTERN BUILDING UNDERLAY CARDS
UNIFIX BLANK UNDERLAY CARDS
UNIFIX GUMMED SHEETS

Activities:

This activity is best when used with small groups of students.

Display one of the *Pattern Building Underlay Cards* and lead the students in describing the pattern.

"Is there a design that you can see in the pattern card?"

"Are the four corners of the pattern card the same?" "Different?" "If different, how are they different?"

"Do you think there are more blue than red squares? More yellow or green squares?"

Place the pattern card in the *Operational Tray* and cover with the *Grid.* Ask the students to cooperatively follow the pattern and fill the grid with *Unifix Cubes.* Each student can be asked to fill one row with cubes.

After the pattern has been completed, distribute *Blank Underlay Cards* and *Gummed Sheets* to the students. Ask the students to make a copy of the Pattern they have just completed. After the students have completed their patterns, ask, *"Which color covers the most squares?"*

To verify the students' estimates of number, ask the students to remove the Unifix Cubes from the Operational Grid and Tray and separate them by color. Ask the students to put all the red cubes together, all the blue cubes together, etc. in trains or towers. Without counting, ask which train has the most cubes. Then ask the students to take the cubes apart and put back into piles. *"Which pile has the most cubes?"*

If the students seem uncertain of this operation, repeat this activity with other Pattern Building Underlay Cards until students appear confident in their understanding that the number of units does not change when the appearance changes.

This activity can be further simplified by asking the students to cover only the yellow and green areas of the pattern or only the red and blue areas and follow the directions above.

2 Building Graphs

Graphs are representations of numerical information, ideas and relations. Collecting, organizing, describing, interpreting and displaying this information in a pictorial manner helps us to communicate better. Graphing also gives us another dimension for transmitting meaning and is important in terms of making intelligent decisions and solving problems.

In addition to gathering data and building graphs, asking questions and comparing and interpreting the results are also important. When students collect, record and compare data, they should be asked such questions as: *"What patterns do you see from the data?" "Were you surprised with the results of your graph patterns?"* etc.

The most frequently used graphs are:

- **Picture Graphs** in which pictures represent data.
- **Bar Graphs** in which bars represent percentages and other data.
- **Line Graphs** in which lines are drawn on grids.
- **Circle Graphs** or **Pie Charts** in which a circle represents the whole and the wedges represent parts or percentages of the whole.

Unifix Cubes are excellent tools for teaching bar graphing. They come in ten colors and can represent many variables. They are uniform in size, representing a unit or a given number of units, and they can be snapped together to display the total for a given category. The following lessons illustrate how Unifix Cubes and ancillary materials can be used to demonstrate building bar graphs.

Cubes

Blank Underlay Cards

Rod Stamps

Operational Grid and Tray

10 x 10 Number Tray

Gummed Sheets

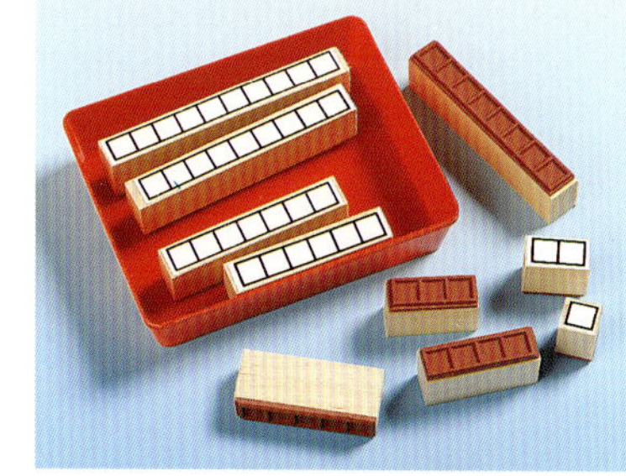

Rod Stamps

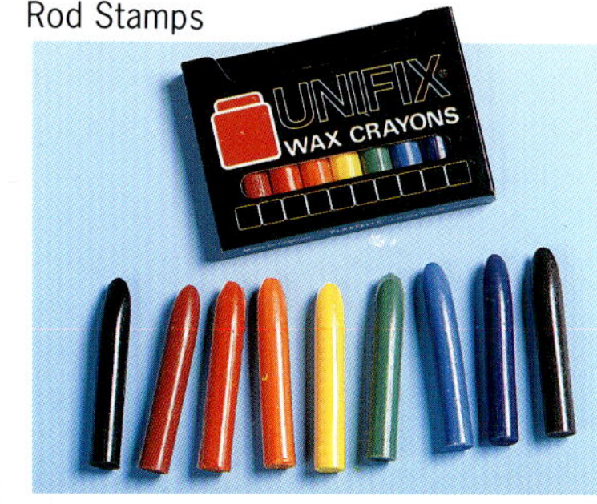

Wax Crayons

Lesson 1

Objective:

After collecting data and constructing a vertical bar graph with Unifix Cubes, students will understand that there is a base or data point for each of the bars and that the graph is built by units to the value axis.

Materials:

UNIFIX CUBES
UNIFIX 10 X 10 NUMBER TRAY
UNIFIX GUMMED SHEETS
UNIFIX BLANK UNDERLAY CARDS

Activities:

Tell the students that they will be making a graph about favorite zoo animals. Ask the students to name ten zoo animals and assign each a number from 1 to 10. Write the list on the chalkboard.

Label the base of the *10 x 10 Number Tray* with the numbers 1 to 10. Then ask each student to name his or her favorite animal and place a *Unifix Cube* in the corresponding groove on the Number Tray.

Discuss the resulting bar graph. *"Which animal is best liked?" "Which animal is least liked?" "How many more students like penguins than like camels?"*

As an extended activity, ask the students to make a permanent copy of the bar graph with *Unifix Blank Underlay Cards* and *Gummed Sheets* and label each column with the number of each zoo animal.

Lesson 2

Objective:

Students will collect data and construct a horizontal bar graph using several different media.

Materials:

UNIFIX CUBES
UNIFIX OPERATIONAL GRIDS AND TRAYS
UNIFIX BLANK UNDERLAY CARDS
UNIFIX GUMMED SHEETS

Activities:

Assign teams of students to work together collecting data and constructing horizontal bar graphs using *Unifix Cubes* and the *Operational Grid and Tray*.

Each group can research a different subject of up to ten items such as favorite pets or favorite foods. After students tally the results of their inquiries, ask them to list the names of the subjects on a piece of paper and place the list at the left side of the Operational Grid and Tray.

A Unifix Cube should be placed in the grid going from left to right to represent one unit of the subject researched.

After all teams have completed their bar graphs, ask a member of each team to explain the graph to the class. As an extended activity, each member of each team can construct a copy of the bar graph using *Unifix Blank Underlay Cards* and *Unifix Gummed Sheets*.

Lesson 3

Objective:

Students will create bar graphs using rubber stamps and crayons.

Materials:

UNIFIX ROD STAMPS - ACTUAL CUBE SIZE
UNIFIX WAX CRAYONS

Activities:

Distribute *Unifix Rod Stamps, Wax Crayons* and sheets of paper to groups of students.

Begin this activity by asking each student to name his or her favorite color.

Tally this data on the chalkboard and ask the students to create a graph using this data. Using the Unifix Rod Stamps, students create bar graphs representing the colors and number the side of their papers from 1 to 10.

Students then use the Wax Crayons to complete the graph and then discuss what each space on each bar represents.

What is the favorite class color?

Additional graphing subjects:
- *The month with the most birthdays*
- *Daily temperature for two weeks*
- *Favorite television programs*

3 Counting and Place Value to Twenty

In beginning numeration it is critical that students gain an understanding of the numbers zero through twenty.

Students need to use many mathematical models with a variety of activities to develop a real understanding of the numbers to twenty. It is important that students have a clear idea of the meaning of numbers prior to beginning to work with the basic addition and subtraction facts.

The following lesson plans illustrate ways in which these numbers can be made meaningful for your students.

1 - 10 Stair

One-Ten and Ones Tray

1 - 5 Stair

Number Indicators

Number Cards

Counting Ladder

Value Cards

Inset Pattern Boards

My First Unifix Counting Book

Counting and Place Value to Twenty

Lesson 1

Objective:

Students will gain an understanding of the numbers one through five.

Materials:

UNIFIX CUBES
UNIFIX VALUE CARDS 1 - 5
UNIFIX 1 - 5 STAIR
UNIFIX INSET PATTERN BOARDS - DOMINO AND TWO'S PATTERNS
MY FIRST UNIFIX COUNTING BOOK

Activities

"Name the Number" is a simple activity that will help your students identify numbers without counting and then relate them to the written representations of the numbers.

To play, place two, three, four or five *Unifix Cubes* in your hand or in a bag. Toss the cubes onto the table and call on a student to tell the number of cubes without counting.

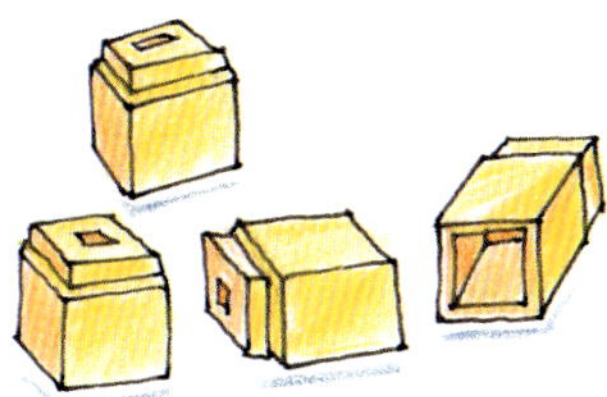

Then, for example, if the set is four cubes and the student names the correct number, ask the student to lock the four cubes together and place the tower of cubes above the number 4 on the 1 - 5 side of the *Unifix Value Card*.

Continue with the remaining numbers of one, two, three and five.

As a variation of "Name the Number," use *My First Unifix Counting Book* to help the students make the connection of the number of cubes, the number and the number name.

After a student names the number of cubes tossed on the table, ask the student to find the corrrect page in the book and to place the cubes on the page.

Then the same number of cubes can be connected into a bar and placed on the page, demonstrating equivalence.

Use the Unifix 1 - 5 Stair to further extend the activity.

Ask the students to place the tower of cubes they made in "Name the Number" below the correct number on the stair.

When the 1 - 5 Stair is completed, ask questions such as:

"Which number is one more than 4? One more than 2?"

As an additional activity, students can work in pairs taking turns in placing Unifix Cubes in the *Inset Pattern Boards* using both *Domino and Two's Patterns* for the numbers one through five.

After the students count the cubes and say the number, they should find the correct number card and place the card over the numeral on the board.

Lesson 2

Objective:

Students will gain an understanding of the numbers six through ten.

Materials:

UNIFIX CUBES
UNIFIX VALUE CARDS 6 - 10
UNIFIX NUMBER INDICATORS
UNIFIX 1 - 10 VALUE BOATS
UNIFIX 1 - 10 STAIRS

Activity:

Demonstrate and explain the activity, **"What is One More?"** The activity is played by individual students or students working in groups.

Give the students a quantity of *Unifix Cubes* and ask them to join the cubes together to make towers of cubes starting with six cubes and building up each tower to be *"one more"* than the previous tower until they have made a tower of ten cubes.

The towers of cubes are then placed on the *Unifix Value Cards,* matching the numbers on the card.

When finished, the students place *Number Indicators* on top of each tower.

Ask questions about the towers such as: *"What number is one more than seven? One more than five? One more than nine?"*

The *1 - 10 Stair* can also be used to demonstrate the "one more" concept. Students remove the Number Indicators from their towers of cubes and find the correct stair in which that tower will fit.

As a further extension of this activity, the towers of cubes can be placed in the *1 - 10 Value Boats* and placed in order from one to ten.

Ask questions such as:

"Which two Value Boats could match the length of the Value Boat for seven?"

"Is there more than one answer?"

"Which two Value Boats could be used to make the numbers six, eight, nine and ten?"

Lesson 3

Objective:

Students will gain an understanding of the meaning of numbers to 19.

Materials:

UNIFIX CUBES
UNIFIX ONE-TEN AND ONES TRAYS
UNIFIX NUMBER CARDS
A DIE

Activities:

Explain and demonstrate the activity **"Make 19"** for your students.

The game requires two players and each player has a *One-Ten and Ones Tray* and *Unifix Cubes.*

The players take turns rolling the die and placing cubes in the tray.

If, for example, a player rolls a 5, the player places five cubes in the ones groove of the tray.

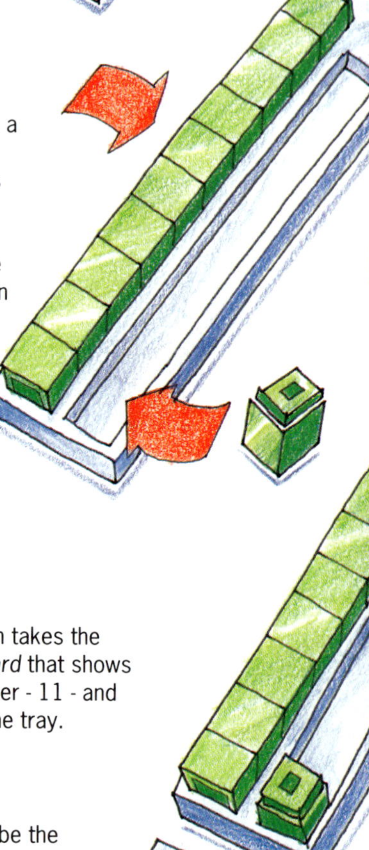

The next time that player has a turn, he or she might roll a 6. The player takes 6 Unifix Cubes and also takes out the 5 cubes from the tray, and then places 10 cubes in the tens part of the tray and the one leftover cube in the ones groove of the tray.

The player then takes the *Unifix Number Card* that shows the correct number - 11 - and places it below the tray.

The object of the game is to be the first to make exactly 19 .

If a player rolls a number that would make a number greater than 19, the player loses that turn and must wait for the next turn to roll the die.

Lesson 4

Objective:

Students will gain an understanding of the numbers 11 through 20.

Materials:

UNIFIX CUBES

UNIFIX ONE-TEN AND ONES TRAYS

UNIFIX NUMBER CARDS

BLANK SPINNER

Activities:

Using a blank spinner, write the numbers one through nine or make your own spinner using a piece of heavy board, a paper clip and a brad.

Demonstrate and explain the game **"High Spinner"** for your students.

Distribute 19 *Unifix Cubes* and a *One-Ten and Ones Tray* to each player. The game is played by two students.

The first player spins the spinner, takes enough cubes to match the number on the spinner and places them in his or her tray.

For example, if the spinner stops on the number nine, nine cubes are placed in the ones part of the tray.

The second player then takes his or her turn.

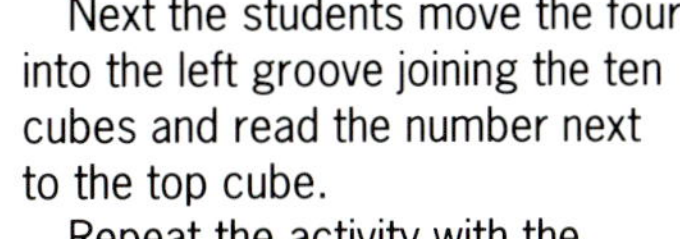

The first player takes a second turn. If he or she should spin the number five, he or she takes five more cubes and moves the nine cubes from the ones groove into the ten groove adding one more cube to make a ten and then puts the remaining four cubes in the ones part of the tray.

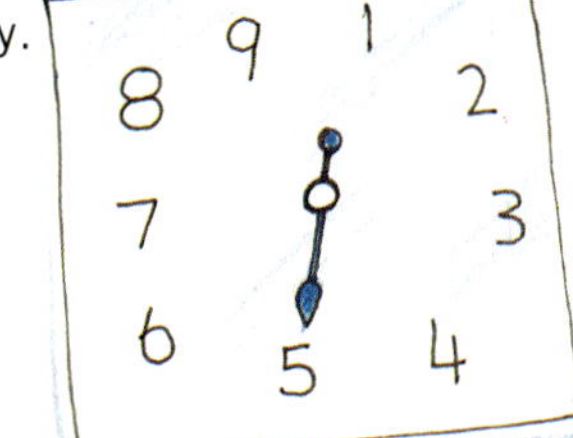

The player then finds the correct *Unifix Number Card* - 14 - and places it below the tray. The second player then takes his or her turn and follows the same procedure as player one. The players record the numbers in each tray and remove the cubes. The winner of that round is the one with the highest number. The game is repeated and the grand winner is the one with the most rounds out of five games.

Lesson 5

Objective:

Students will gain a greater understanding of teen numbers and see that teen numbers are one ten plus ones.

Materials:

UNIFIX CUBES

UNIFIX NUMBER CARDS

UNIFIX COUNTING LADDERS

Activities:

To further reinforce the teen numbers and to demonstrate that teen numbers are always one ten plus ones, students can use *Unifix Cubes, Number Cards* and the *Counting Ladder.*

Each student should have a supply of Unifix Cubes, Number Cards and a Counting Ladder.

Tell the students that you are going to say a number and they are to represent that number with cubes on the Counting Ladder.

For example, ask the children to show the number 14 by first finding the correct Number Card and placing it under the Counting Ladder.

Then they should join ten cubes and place them in the left groove and place the four remaining cubes in the right groove.

Next the students move the four cubes into the left groove joining the ten cubes and read the number next to the top cube.

Repeat the activity with the other numbers until students are confident with the process of creating teen numbers.

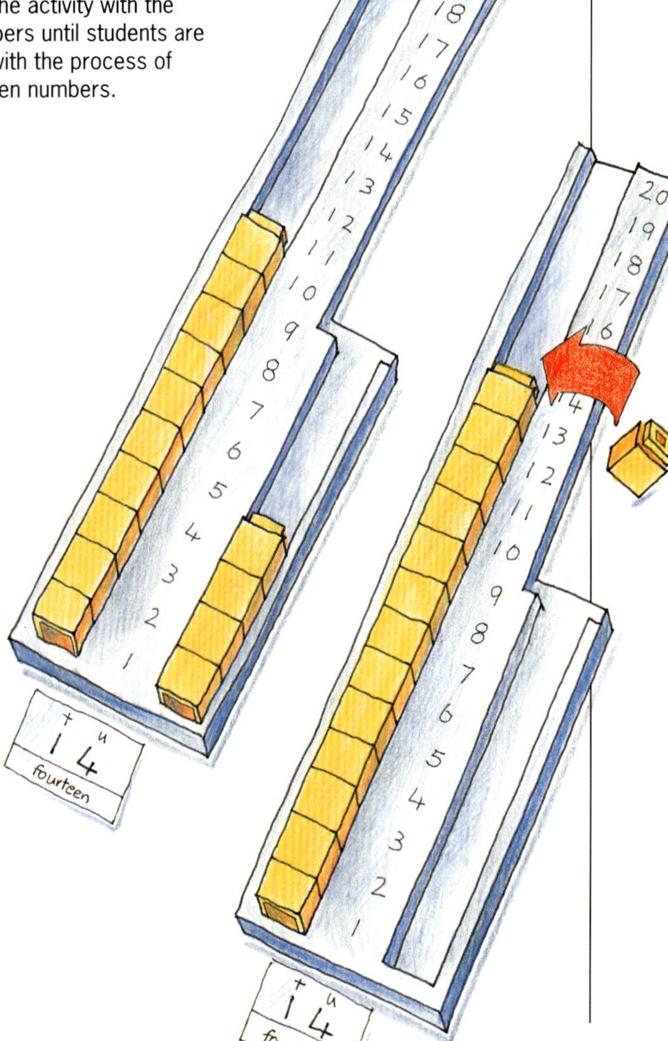

4 Meaning of Addition

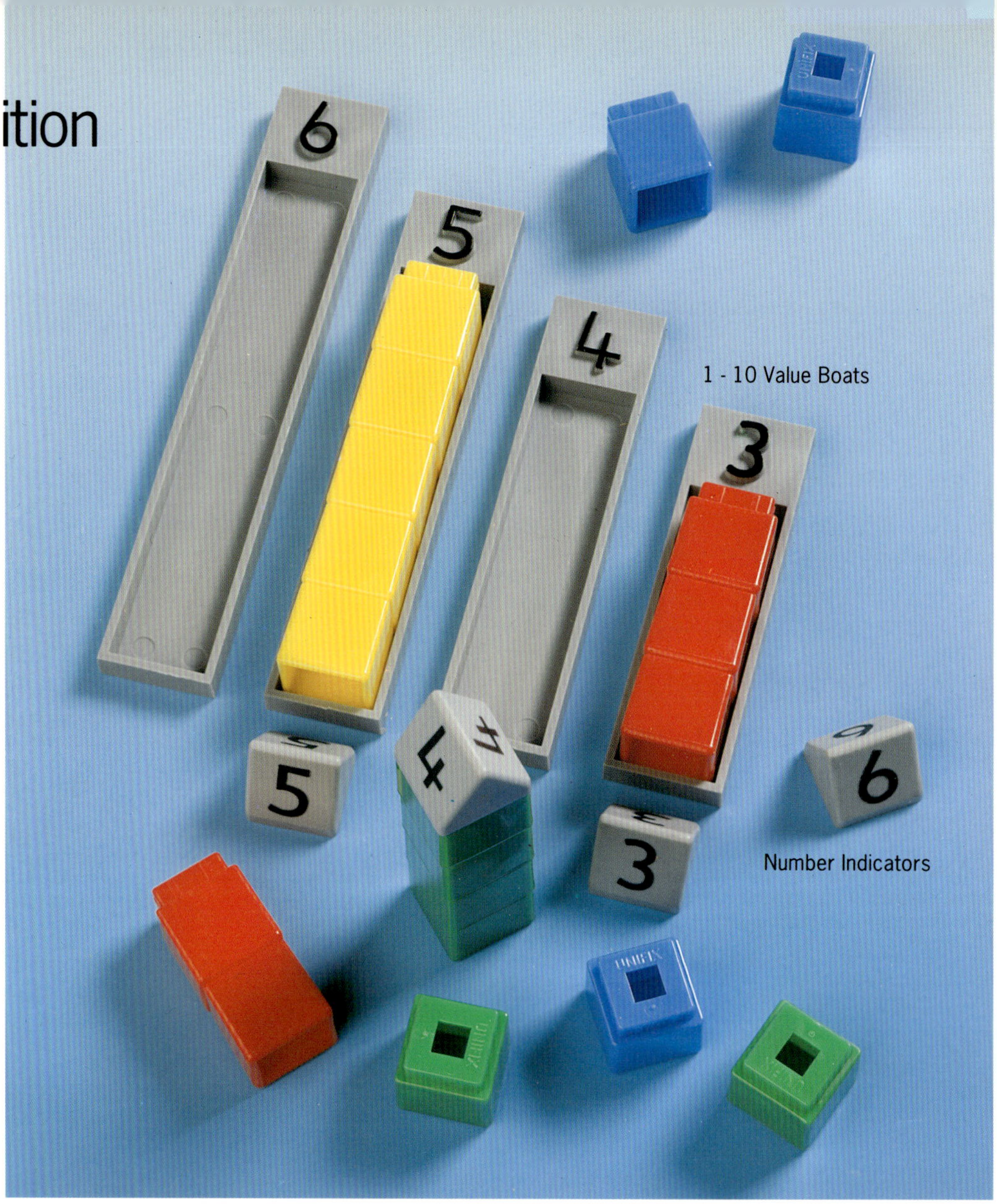

The operation of addition can be described as *"part plus part equals the whole."*

Addition is a binary operation in which you have two input and output.

You can add only two numbers at one time and when there are more than two numbers to be added, the numbers need to be grouped. Two numbers are added, then the next number is added to the sum of the first two numbers.

Example: $(2+3)+5=$

or $2+(3+5)=$

To understand the meaning of addition, students need to use materials in teacher-directed activities that will demonstrate the concepts. There is a need to physically put together two sets of objects to make the larger set.

1 - 10 Value Boats

Number Indicators

Blackline Master #1

Note:

Mathematics manipulatives can create bridges or connectors from the concrete world of real things to the abstract, mathematical symbolic world.

Connectors are all materials, games and activities that teachers use to help students "see" into and become competent in the world of mathematics.

See Chapter 16, page 57 for a graphic representation of the concept of connectors.

Meaning of Addition

Lesson 1

Objective:

Students will gain an understanding of addition as part plus part equals the whole.

Materials:

UNIFIX CUBES
BLACKLINE MASTER #1

Activities:

Make enough copies of *Blackline Master #1* for the students who will be participating in the activity.

Distribute a quantity of red and green *Unifix Cubes* and a copy of Blackline Master #1 to each student.

Tell a story such as: *"The balloon man has four red balloons and three green balloons. How many balloons does he have all together?"*

Ask the students to place four red cubes in one rectangle of the blackline master and three green cubes in the other rectangle. Then ask them to pull down all of the cubes to the bottom rectangle.

As the students pull down the cubes, they can say, *"Part plus part equals the whole."* Now ask the students to count the number of cubes in the large rectangle.

The activity can be recorded on the chalkboard as 4 + 3 = 7.

Repeat the activity with different stories that involve numbers with the whole no greater than ten. Ask the students to make up stories and share them with the class or their group.

Additional activities using Blackline Master #1 can be found in Chapter 16.

Lesson 2

Objective:

Students will understand that a given number can be made up of different parts.

Materials:

UNIFIX CUBES
UNIFIX 1 - 10 VALUE BOATS
UNIFIX NUMBER INDICATORS

Activities:

Tell the students that they will be doing an activity called **"Many Names for a Number"** and they will work together in groups of two or three.

Give each group a *Value Boat*, a *Number Indicator* and a quantity of *Unifix Cubes.*

Ask the students to fill the Value Boat with two different colored Unifix Cubes. For example, if a group has the 5 Value Boat, the students could put three green cubes and two red cubes in the boat.

The "name" of the number 5 Value Boat would be *"three cubes and two cubes are five cubes."*

The students should then take the cubes out of the Value Boats, lock them together to form a tower and place the Number Indicator 5 on top of the tower.

One of the students in the group should record on paper 3 + 2 = 5 and 2 + 3 = 5.

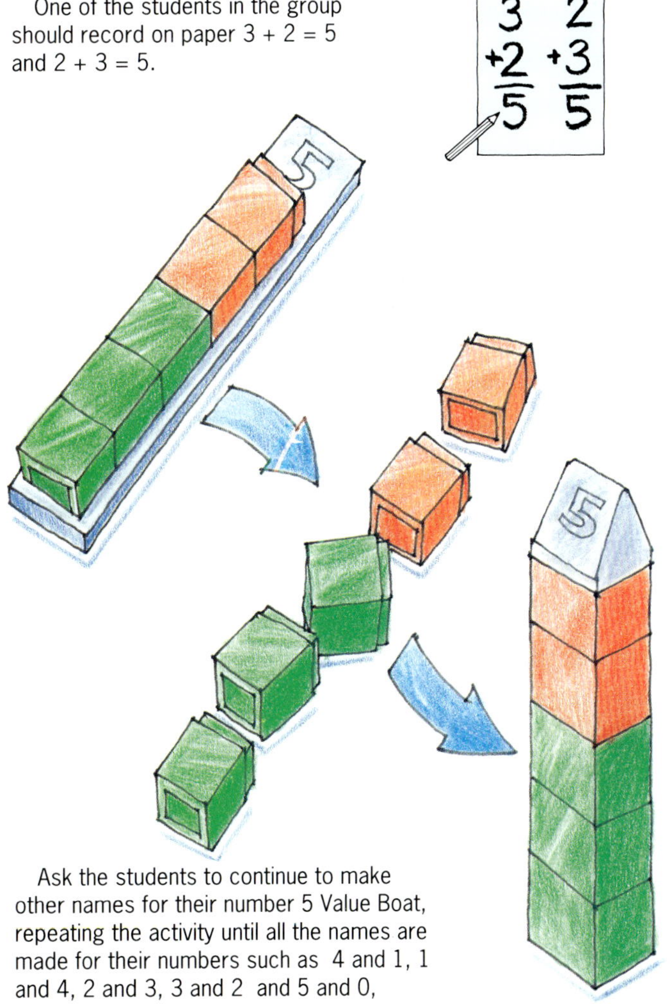

Ask the students to continue to make other names for their number 5 Value Boat, repeating the activity until all the names are made for their numbers such as 4 and 1, 1 and 4, 2 and 3, 3 and 2 and 5 and 0, 0 and 5.

These activities should be extended using the 1-10 Value Boats.

UNIFIX

5 Addition Facts

Addition facts are learned after the students have gained an understanding of the meaning of addition. Addition facts are the basic groundwork needed before students learn to add large numbers.

In the beginning, students need models for the addition facts and should use manipulative materials to create those models and then record the entire fact. The best way to help your students memorize the basic facts is to ask them to write the whole fact. Recording only the sum on paper does not help achieve mastery of the facts.

It is also important that students understand the commutative property of addition which states that you can switch the addends and the sum remains the same (3 + 2 = 5 and 2 + 3 = 5). The commutative property helps students learn the basic facts.

Calculators should not be used for the basic facts except in extreme cases of disability.

Addition Facts

Lesson 1

Objective:

Students will join sets of cubes to model the addition facts with sums to ten.

Materials:

UNIFIX CUBES

PAPER AND PENCILS

Activities:

In this activity, the students are challenged to join Unifix Cubes and then to put together sets of cubes to represent the addition facts. Students should work together in groups of three or four students.

Distribute a quantity of Unifix Cubes to each group of students, along with paper and pencil to record the activity.

Ask each group of students to make trains of 2, 3, 4 and 5 cubes using different colored cubes for each train. For example, all 2 trains use red cubes, all 3 trains use green cubes, etc. The activity is to have the students join two different colored trains together and then record the action taken.

Ask, for example, *"How many ways are there to make a train of six cubes?"*

Encourage the students to construct trains of 2 and 4, 3 and 3, 4 and 2, 5 and 1, 1 and 5 until they have constructed all of the addition facts for six.

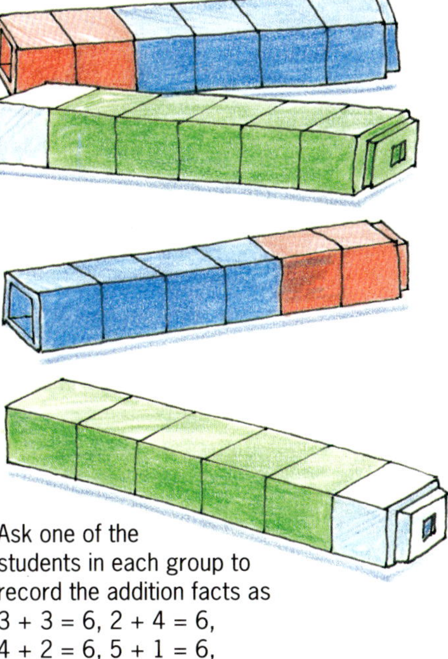

Ask one of the students in each group to record the addition facts as 3 + 3 = 6, 2 + 4 = 6, 4 + 2 = 6, 5 + 1 = 6, 1 + 5 = 6.

After each group of students has completed the addition facts, ask them to compare their results.

4 2
+2 +4
6 6

Lesson 2

Objective:

Given a number, students will create models of the addition facts of that number using Unifix Rod Stamps and then record the addition facts.

Materials:

UNIFIX ROD STAMPS - ACTUAL SIZE

PAPER AND PENCILS

UNIFIX WAX CRAYONS

Activities:

Working with a group of two or three students, give the students a number such as 7 and ask them to use the *Rod Stamps* to make pictures of that number.

Students choose two stamps such as 4 and 3 and join the impressions to form a picture of 7 and write the addition fact 4 + 3 = 7 beneath the picture.

Students may also color the four cubes one color and the three cubes another color.

Students should continue making 7 combinations until they have samples of all of the addition facts for that number.

Students can share their work with the other students so they all can see the facts for the other sums.

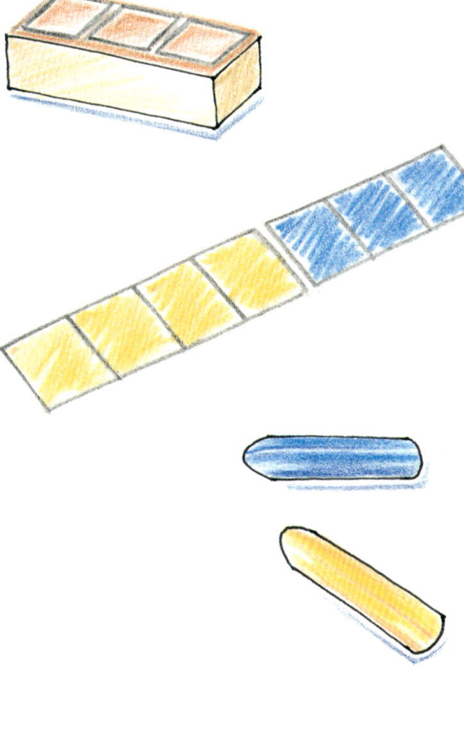

Lesson 3

Objective:

Students will learn to trade ones to make a ten to find the sum of the facts with sums 11 to 20.

Materials:

UNIFIX CUBES

UNIFIX COUNTING LADDERS

Activities:

Distribute a quantity of *Unifix Cubes* and a *Counting Ladder* to each group of students.

Write a fact such as 8 + 7 = ? on the chalkboard.

Ask the students to take eight cubes and place them in the left hand column and then place seven cubes in the right side column. Ask, *"How many cubes must we move to have ten in the left hand column?"*

Students should then move two cubes into that column leaving five in the right hand column.

The sum of 8 + 7 is one ten and five ones or 15. This is verified by the students placing the five ones on top of the ten cubes and reading the number 15 on the ladder.

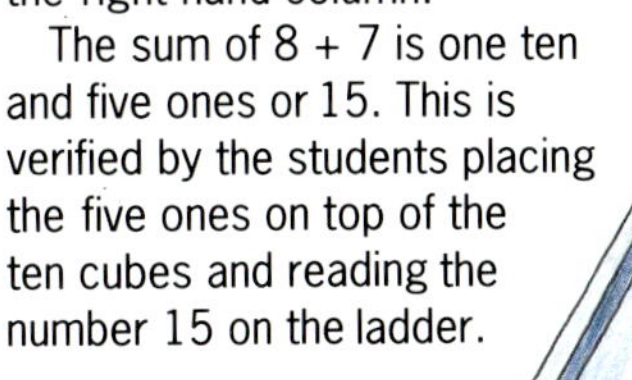

Repeat this activity with other facts such as 7 + 6, 8 + 4, 9 + 7, etc.

The number of cubes for the greatest addend is always placed in the left column of the tray. After each activity is completed, ask the students to write the whole fact.

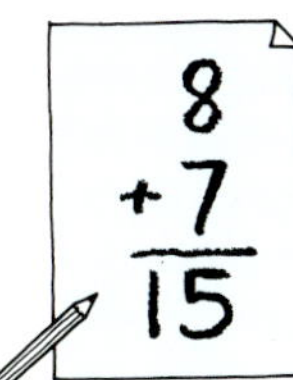

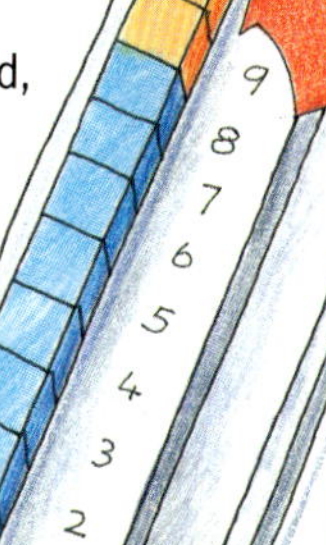

6 Meaning of Subtraction and Subtraction Facts

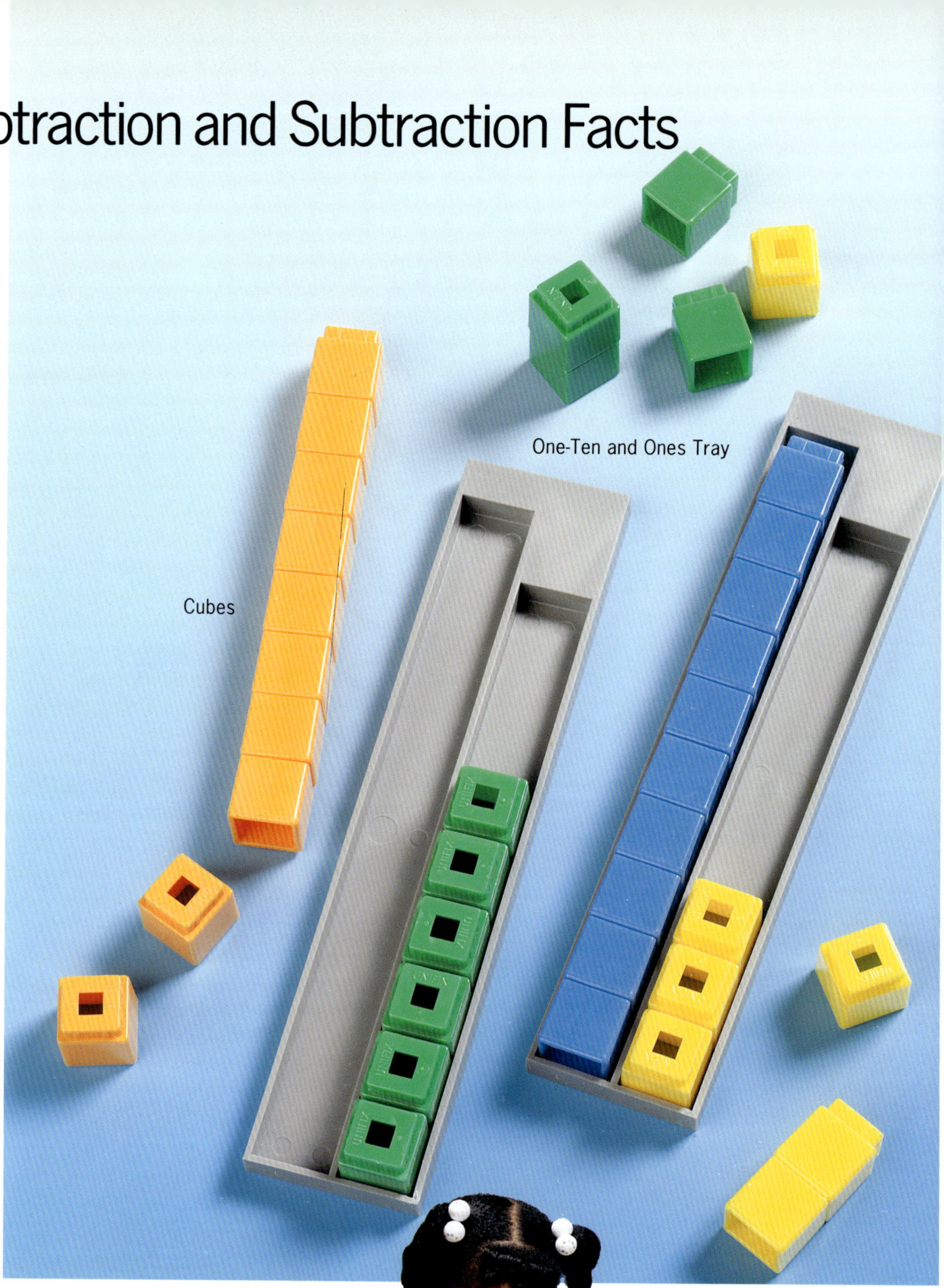

The operation of subtraction is the inverse operation of addition, and an inverse operation is always more difficult to learn. There are various models of subtraction. The following lessons offer two of the main models.

The least abstract is the "take away" model and the more abstract model is the "add on" model.

These two models need to be taught with many activities in which students are involved with manipulative materials. Building the concept of subtraction is necessary before students operate with numbers.

After your students have learned the basic addition facts and have explored the meaning of subtraction using manipulatives, they are ready to learn the basic subtraction facts. It is essential that students see that the addition and subtraction facts are related.

Activities using manipulative materials will help students develop models to understand the facts. Writing and saying the whole fact as the action is performed will help students remember the facts.

Note:

Mathematics manipulatives can create bridges or connectors from the concrete world of real things to the abstract, mathematical symbolic world. Connectors are all materials, games and activities that teachers use to help students "see" into and become competent in the world of mathematics.

See Chapter 16, page 57, for a graphic representation of the concept of connectors.

Blackline Master #1

One-Ten and Ones Tray

Cubes

Meaning of Subtraction and Subtraction Facts

Lesson 1

Objective:

Students will gain an understanding of the "take away" model of subtraction.

Materials:

UNIFIX CUBES

BLACKLINE MASTER #1

Activities:

Duplicate enough copies of *Blackline Master #1* for your students and distribute a copy of the worksheet and a quantity of *Unifix Cubes* to each student.

Write a subtraction problem such as 8 - 5 = ? on the chalkboard.

Ask the students to place eight cubes in the large rectangle at the top of their worksheets.

Then five cubes are taken away from the eight cubes and placed in the right bottom section of the worksheet.

Finally, ask the students to pull down the remaining three cubes into the left bottom section.

On a separate piece of paper, students record 8 - 5 = 3.

Point out to the students that their worksheets also demonstrate that 5 + 3 = 8.

Another model of "take away" subtraction can be demonstrated using Unifix Cubes in "trains".

Write a subtraction problem such as 10 - 4 = ? on the chalkboard.

Ask the students to connect ten Unifix Cubes to create a ten-cube "train." Then they are to take away four cubes to show that the missing part is "6."The students then record 10 - 4 = 6.

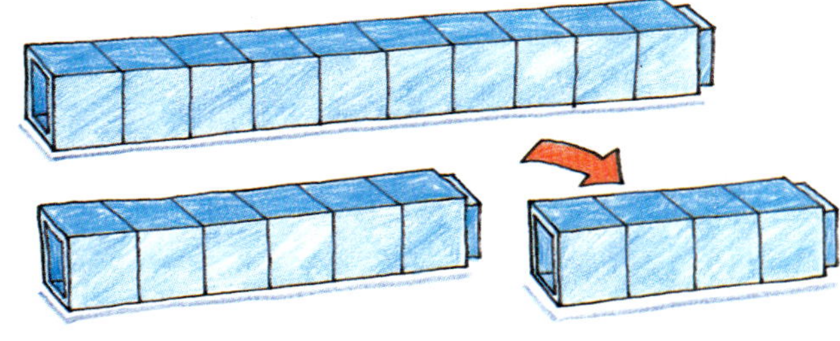

A variety of activities can be developed using the Unifix Cube "train" model.

Students can be given a subtraction problem such as 13 - 5 = ? They first model 13 by making a train of 13 cubes. Below the train of 13 cubes is placed a train of five cubes.

Ask the students: *"How many cubes must you add on to the five cubes to make 13 cubes?"*

The answer is the missing part of the problem. Students should then record: 13 - 5 = 8.

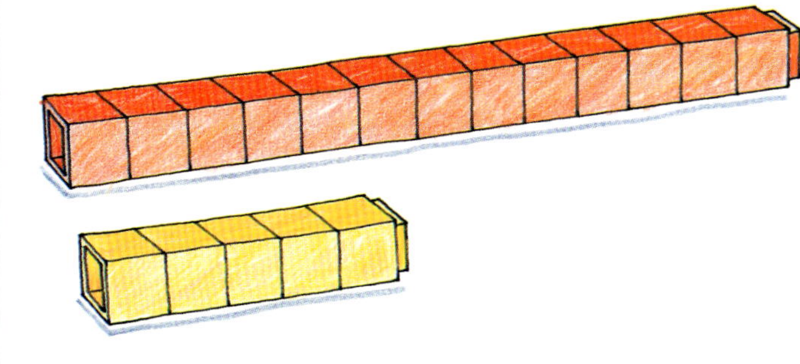

Lesson 2

Objective:

Students will gain an understanding of the subtraction facts through the sum of ten.

Materials:

UNIFIX CUBES

Activities:

Distribute a quantity of *Unifix Cubes* to each student.

In this activity, students are to demonstrate the ways to make a given sum and then find all the subtraction facts with that given sum.

For example, the student connects seven Unifix Cubes to make a seven-cube "train."

Then he or she takes away one, two, three, four, five, six and seven cubes and records each action.

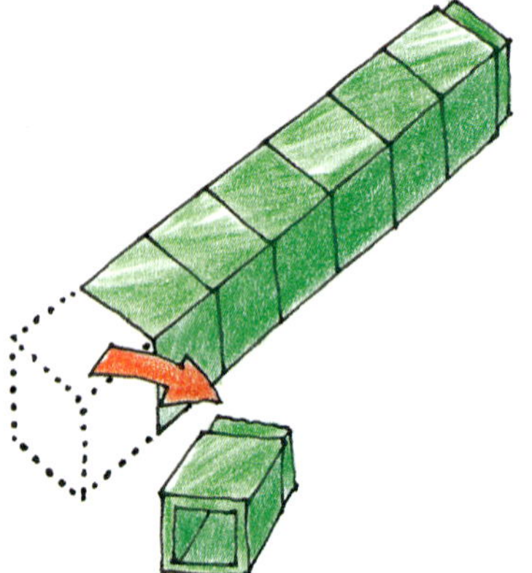

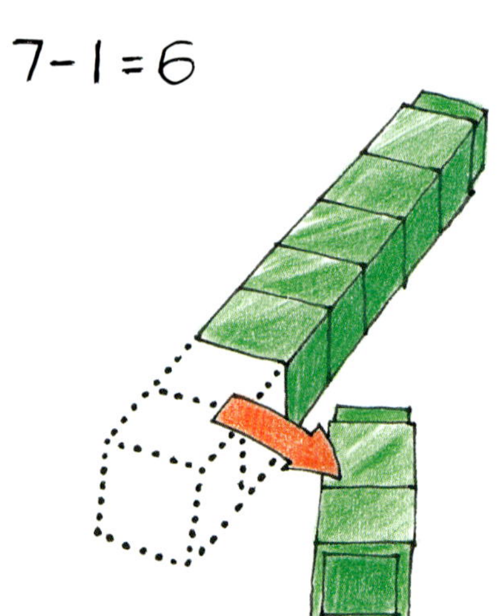

The student then repeats the activity for the other sums of four, five, six, eight, nine and ten.

Lesson 3

Objective:

Students will gain an understanding of the subtraction facts through the sum of 18.

Materials:

UNIFIX CUBES

UNIFIX ONE-TEN AND ONES TRAYS

Activities:

Depending upon the number of *One-Ten and Ones Trays* that are available, students may work individually or in small groups of two or three students in the following activity. Demonstrate the activity for the students, and then allow them to work through other subtraction problems that you present to them.

Distribute a quantity of *Unifix Cubes* and the trays to the students. In this activity, students learn the "tens method" of solving subtraction facts with sums greater than ten.

Write a problem such as
13 - 6 = ? on the chalkboard.

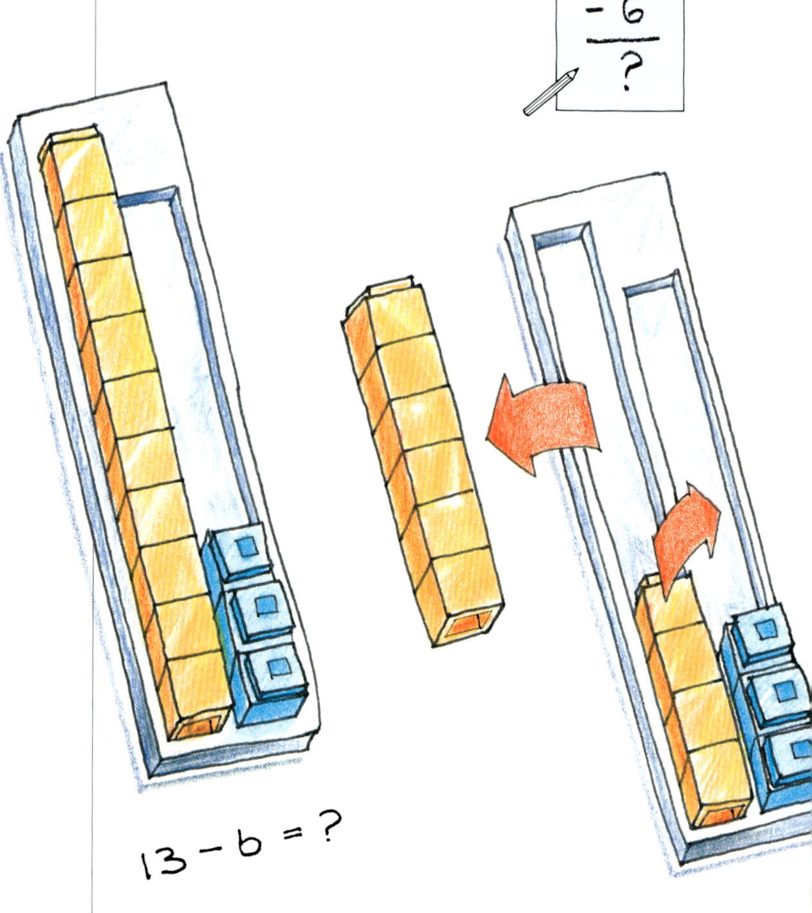

Ask students to count out 13 cubes, connect ten of those cubes into a tower and place the tower in the ten column of the One-Ten and Ones Tray and the remaining three cubes in the ones column. Six cubes are then snapped off the tower of ten. The four remaining cubes are removed from the tens column and placed over the three cubes in the ones column.

Lesson 3

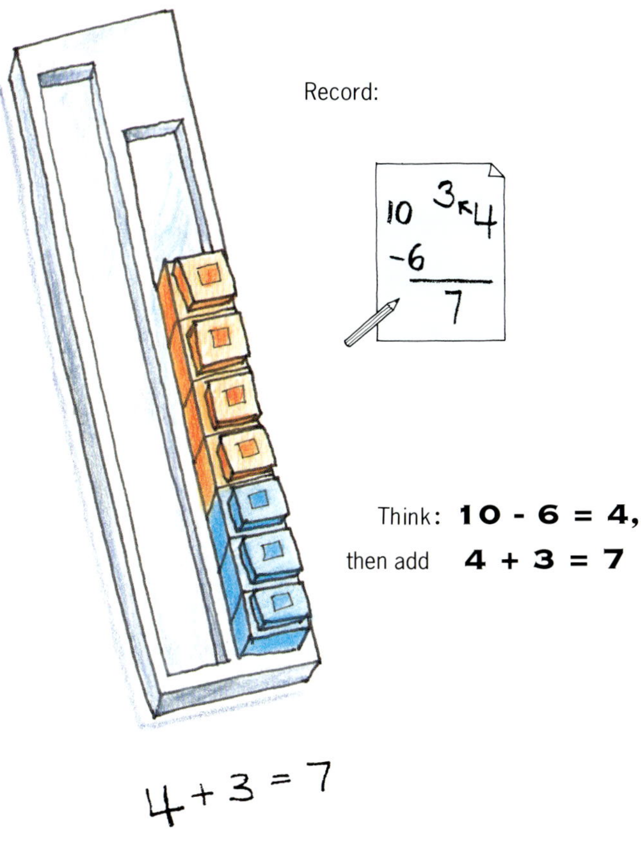

The record of this activity is made by the students in which six is subtracted from ten and then four is added to the three ones. Many activities are needed with other facts to create the model of subtracting from ten and adding to the ones.

After the students have used the Unifix Cubes and the One-Ten and Ones Tray, there is another model they can use called the "hill" method.

This method helps students gain a mental image of subtracting from ten and adding the ones. Demonstrate the subtraction problem 14 - 8 = ? by drawing a hill on the chalkboard and writing 8 at one side of the hill and 14 on the other side of the hill. Write 10 in the middle of the hill.

Ask the students to tell how far it is from 8 to 10. Write 2 on top of the hill between 8 and 10.

Now ask the students how far is it from 10 to 14. Write 4 on top of the hill between 10 and 14.

Now ask the students to add 2 + 4. After the students say 6, record the problem next to the hill on the chalkboard. Give the students at least four additional subtraction problems to solve using the "hill" model.

7 Counting and Place Value Through Hundreds

UNIFIX

One of the most difficult concepts to be learned in elementary mathematics is the understanding of our numeration system that involves place value.

Our system of numeration uses base ten and students must learn that each new place value to the left of the ones is always ten times greater than the place value to the right. To internalize the pattern of base ten, students need many planned experiences with materials that show the pattern of base ten.

It is essential that sufficient time be allotted so students have a good understanding of ones, tens and hundreds before being exposed to larger numbers.

The first step to that understanding is creating a model for numbers up to nineteen and then understanding the value of each digit in two digit numbers.

Hundreds need special attention so students understand that the next place value is made up of ten tens and one hundred ones. Three-digit numbers become difficult since each digit has a separate place value and the value of the number is the sum of the value of each digit.

The following lesson plans illustrate ways in which these concepts can be made concrete for your students.

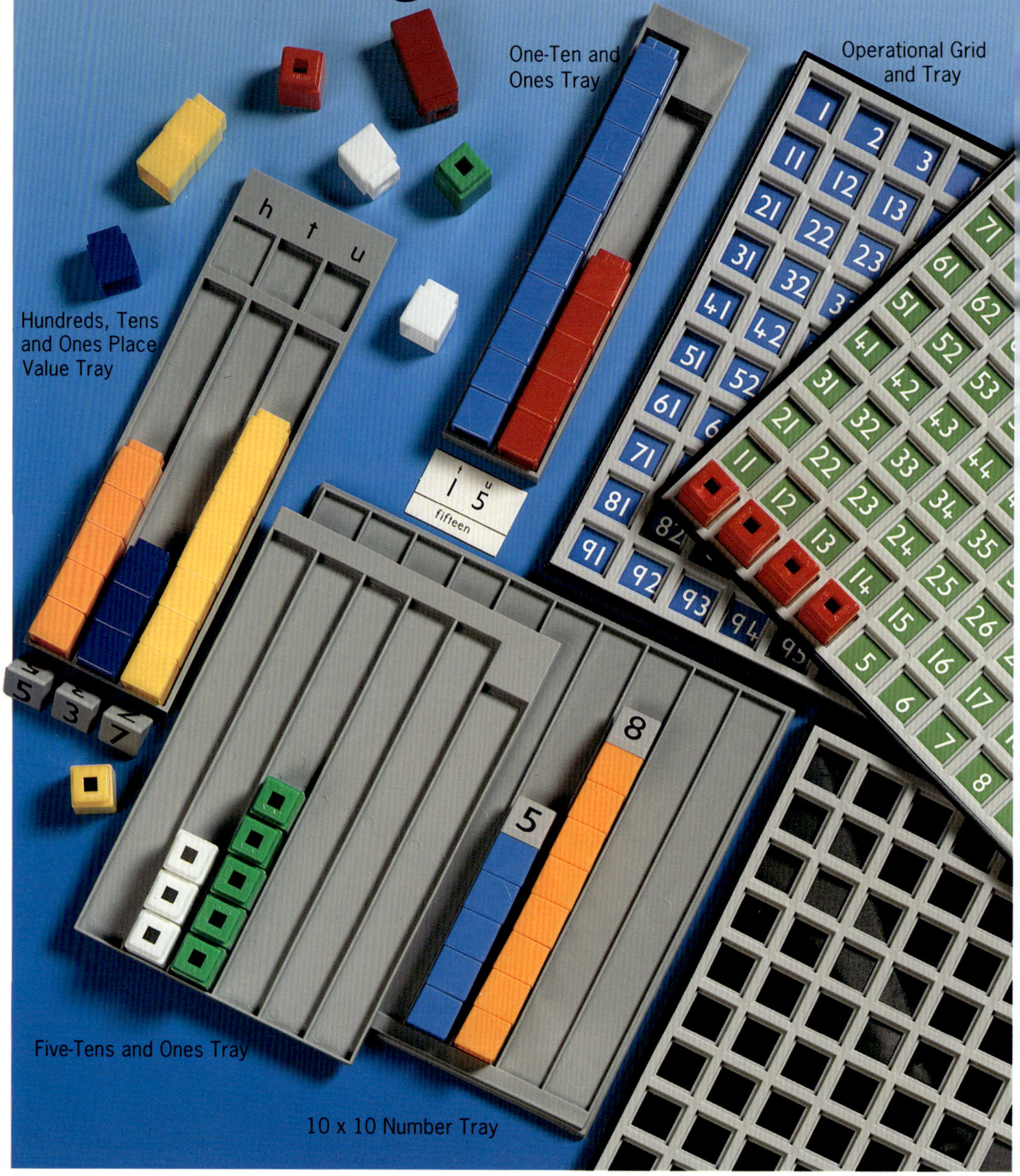

One-Ten and Ones Tray

Operational Grid and Tray

Hundreds, Tens and Ones Place Value Tray

Five-Tens and Ones Tray

10 x 10 Number Tray

100 Track

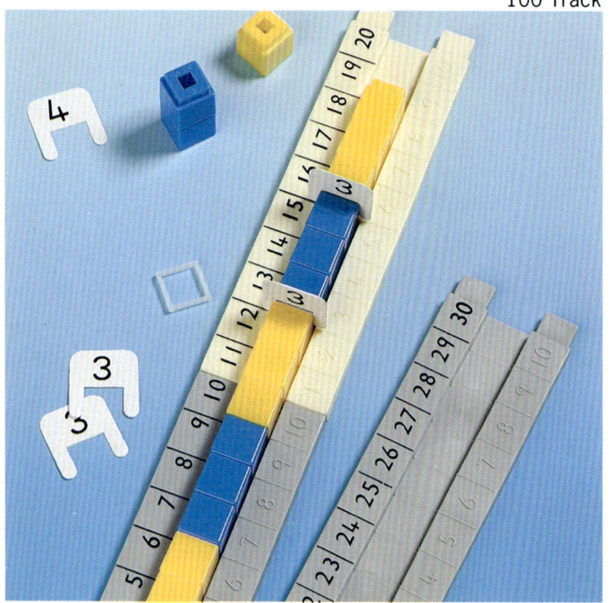

Counting and Place Value Through Hundreds

Lesson 1

Objective:

Students will understand the meaning of two-digit numbers up to 59 and understand which digit stands for tens and which digit stands for ones.

Materials:

UNIFIX CUBES

UNIFIX FIVE-TENS AND ONES TRAY

Activities:

Explain and demonstrate the activity **"Guess the Number"** for your students.

Ask a student to place an arbitrary number of towers of ten and cubes for ones in the *Five-Tens and Ones Tray.*

The other students, who cannot see the tray, ask questions to guess the number that is shown on the tray. The student that placed the cubes in the tray can answer only *"yes"* or *"no."* Students ask questions such as: *"Are there more than two tens? Are there less than seven ones?"* Questions continue until a student guesses the number. The student who guesses the number then puts the cubes in the tray for the next "Guess the Number" game.

"Increase and Decrease" reinforces the place value of the digits.

Ask a student to put the number 27 in the Five-Tens and Ones Tray. Call on a student to increase the number 27 by 20.

The student should place two more towers of ten in the tray. The number is now 47. Ask another student to decrease the number by 3. That student takes three cubes out of the ones side of the tray.

Continue this activity with other numbers until students display confidence in performing the activity.

Lesson 2

Objective:

Students will learn the meaning of a new place value to the left of the tens called a hundred.

Materials:

UNIFIX CUBES

UNIFIX 10 X 10 NUMBER TRAY

UNIFIX 100 TRACK

Activities:

Students need to learn that a hundred is made up of ten tens and one hundred ones.

Students must have the experience of making towers of ten before this activity begins. As towers of ten are placed in the *10 x 10 Number Tray,* the students need to count: *"one ten, two tens, three tens,"* etc. up to nine tens.

Tell the students that 10 tens is a new unit called a hundred. Questions should be asked such as: *"How many tens make a hundred? How many ones make a hundred?"* You also need to show students at this time how to write a hundred with three digits.

If more than one 10 x 10 Number Tray is available, the students can put towers of ten in several trays and begin to count by hundreds.

The following activity prepares students for work with three-digit numbers.

Show the students the 10 x 10 Number Tray filled with ten tens and place a tower of ten to the right of the tray and place one Unifix Cube to the right of that.

Write on the chalkboard the number **111** and point out to the students that the digit "1" to the far left stands for hundred, the middle "1" stands for ten and the "1" to the far right stands for one.

This activity can be repeated showing, for example, one hundred, two tens and three ones.

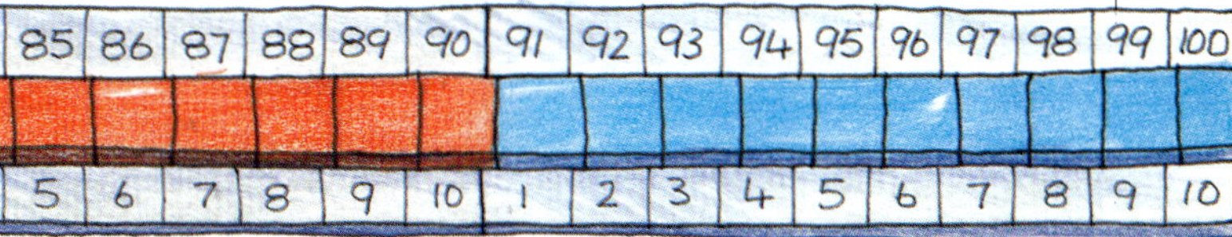

The *100 Track* can also be used to give a linear representation of a hundred. The students can put together the pieces of the 100 Track and then count the number of tens that make up a hundred. The 100 Track also shows that a hundred is made up of 100 ones.

7 Counting and Place Value Through Hundreds

Lesson 3

Objective:

Students will increase their understanding of the meaning of two-digit numbers through 100.

Materials:

UNIFIX OPERATIONAL TRAY AND GRID

UNIFIX UNDERLAY NUMBER CARD

UNIFIX 1 - 100 NUMBER TILES

UNIFIX WINDOW MARKERS

Activities:

For all of the following activities place the *1 - 100 Underlay Number Card* on the *Operational Tray* with the 1 in the top left-hand corner and place the plastic grid on top.

"Follow the Leader" is an activity in which you or the students give directions and the other students respond. Using the tray with the numbers 1 through 100 showing through the grid, one student can say: *"Start at 27 and move two places up. Where do you land?"*

The student that tells the number that is the landing place places a *Window Marker* on that number. After the Window Marker is placed, the student explains how the number was found.

Other questions can be asked such as *"Start at 34, move two places up and one place to the right. Where do you land?"*

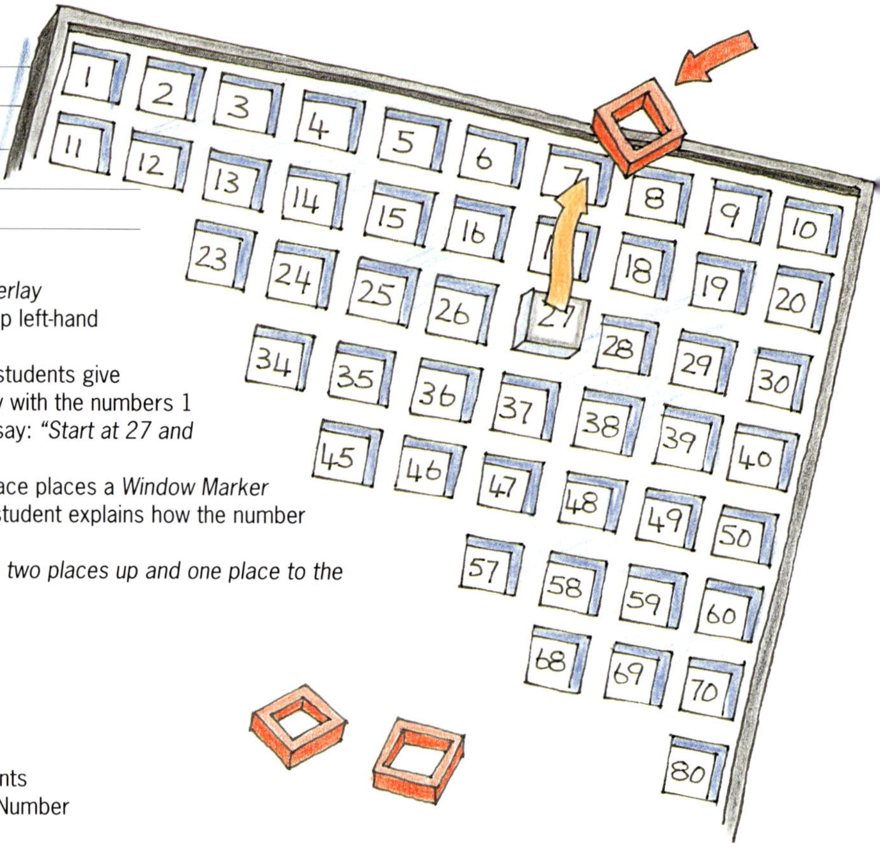

"Find the Numbers" is another activity to keep students looking at all of the two-digit numbers. Students use the Number Tiles to place over the numbers on the Operational Tray.

Ask students to find the numbers which when added the sum is 10, the sum is 9, the sum is 12, etc.

"Many Questions" helps students become aware of the numbers 1 to 100.

Ask a student to pick a number on the 1 - 100 Underlay Number Card. The other students ask questions to find the number. The student that chose the number can answer each question only with *"yes"* or *"no."* Tell the students they should try to find the number by asking the fewest possible questions.

For example, if they ask if the number is greater than 50 they eliminate half of the numbers with that one question. After participating in this activity for awhile, students get better at asking questions. The student that guesses the number can then pick the next number.

Counting and Place Value Through Hundreds

Lesson 4

Objective:

Students will understand the meaning of three-digit numbers, learning the place value of each of the digits.

Materials:

UNIFIX CUBES

UNIFIX HUNDREDS, TENS AND ONES PLACE VALUE TRAY

UNIFIX TENS AND HUNDREDS CUBES

UNIFIX NUMBER INDICATORS

UNIFIX ZERO NUMBER INDICATORS

Activity 1

The activity **"Roll a Number"** involves students in moving objects around and determining the number the objects represent to find the place value of each digit.

Place a number of *Hundred Cubes, Ten Cubes and Unifix Cubes,* representing ones, in a container. Roll the cubes out on the table and ask the students to group all of the Hundreds together, all of the Tens together and all of the Ones together. Then ask a student to demonstrate that number on the Hundreds, Tens and Ones Place Value Tray. For example, if three Hundreds, four Tens and six Ones were rolled, the student places the three Hundred Cubes in the hundreds column, the four Ten Cubes in the tens column and the six Unifix Cubes in the ones column. Ask another student to place the correct Number Indicators below the tray.

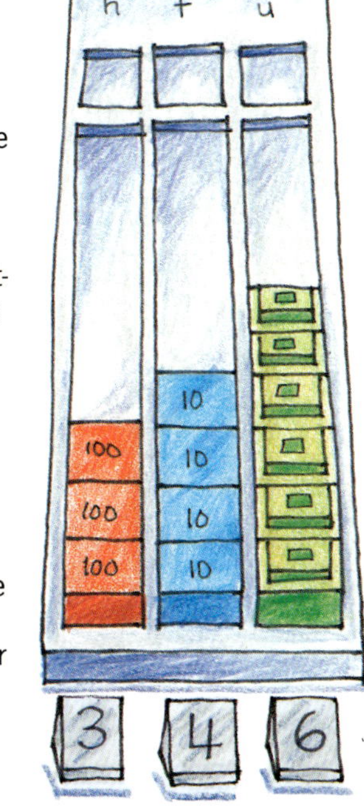

Activity 2

Zero needs special attention in place value of three-digit numbers.

In the activity "Roll a Number" you can take out all of the Tens Cubes and use only Hundred Cubes and Unifix Cubes.

If four 100 Cubes and five Unifix Cubes are rolled, the students will put four cubes in the hundreds column, no cubes in the tens column and five cubes in the ones column.

The number 405 is shown using the Number Indicators and a Zero Indicator. In another activity, use only 100 Cubes. This will give the students an opportunity to indicate on the Hundreds, Tens and Ones Place Value Tray what happens when there is zero in the tens and ones columns. The number should be shown with the Number Indicators and the Zero Indicators.

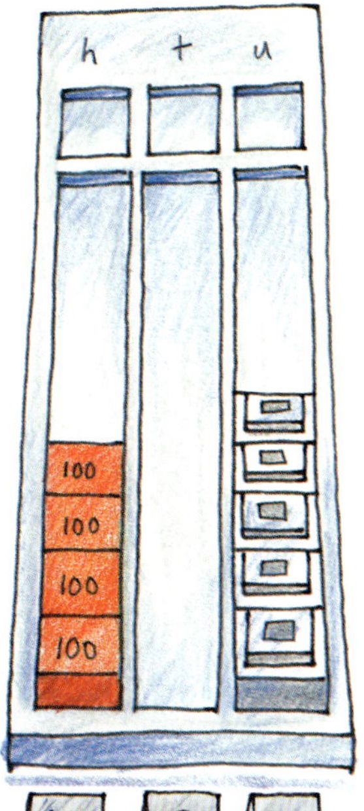

Activity 3

Activities that involve increase and decrease should be done with three-digit numbers to reinforce the place value of each of the digits. For example, ask a student to show the number 347 on the Hundreds, Tens and Ones Place Value Tray. Then ask another student to increase the number by 200. The student should place two more cubes in the hundreds column.

Now ask the students to tell the value of the new number. Then another student can be asked to decrease the number by 30. This student will have to take three cubes from the tens column.

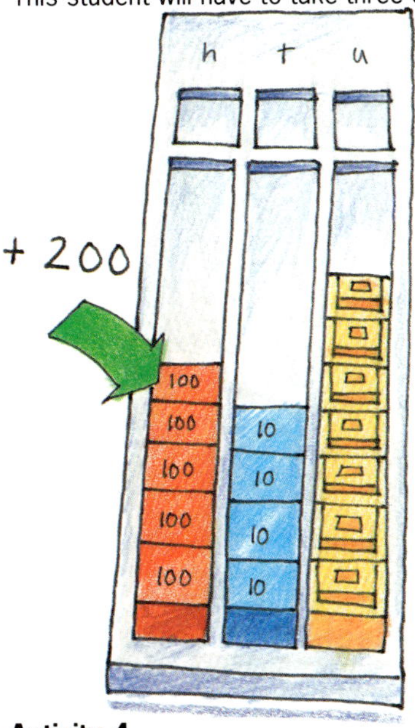

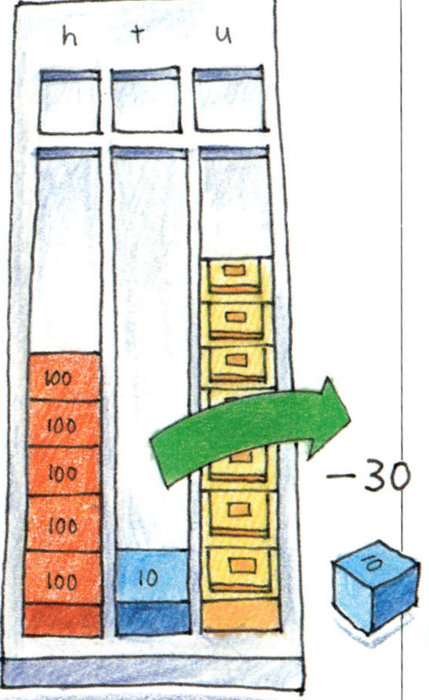

Activity 4

Expanded notation is difficult for many students to understand.

In expanded notation, given the number 348, the students have to write 300 + 40 + 8. Before the students are involved in doing this work in the textbook they need experiences with materials that show the total value of each digit.

Students can be asked to show the number 257 on the Hundreds, Tens and Ones Place Value Tray by putting Unifix Cubes in the columns.

Then ask the students how much the "2" in the number is worth. When the students respond "two hundreds," use the Number Indicators and the Zero Number Indicators to demonstrate the number 200.

Ask how much the "5" is worth. When the students respond "fifty," form the number with the indicators and place below the 200. Then ask how much the "7" is worth. When the students respond "seven," the number 7 is placed below the 50.

This is the vertical form of expanded notation and should be done before the horizontal form is attempted. After repeating the activity with different numbers, the total value of each digit can be placed in the horizontal position, that is, 200 + 50 + 7 = 257.

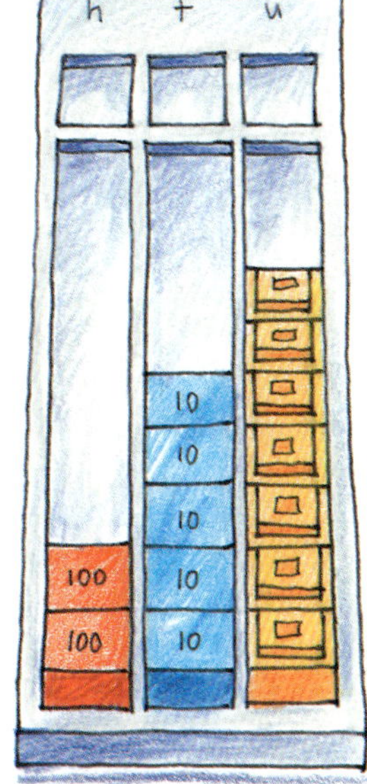

8 Addition of Two- and Three-Digit Numbers

One of the major trouble spots in learning to add numbers with two or more digits is knowing when there is a need to trade units from one place to a larger place in a number. The largest number that can occur in any digit in a number is nine. When more than nine of any unit occurs in addition, a trade must be made for a unit of the next larger place value.

A variety of activities with manipulative materials is needed to help students see the need for trading and how to make the proper trade. By using the proper materials in teaching addition, students can avoid learning the operation by rote.

The following addition activities can be adapted to fit the specific needs of your students.

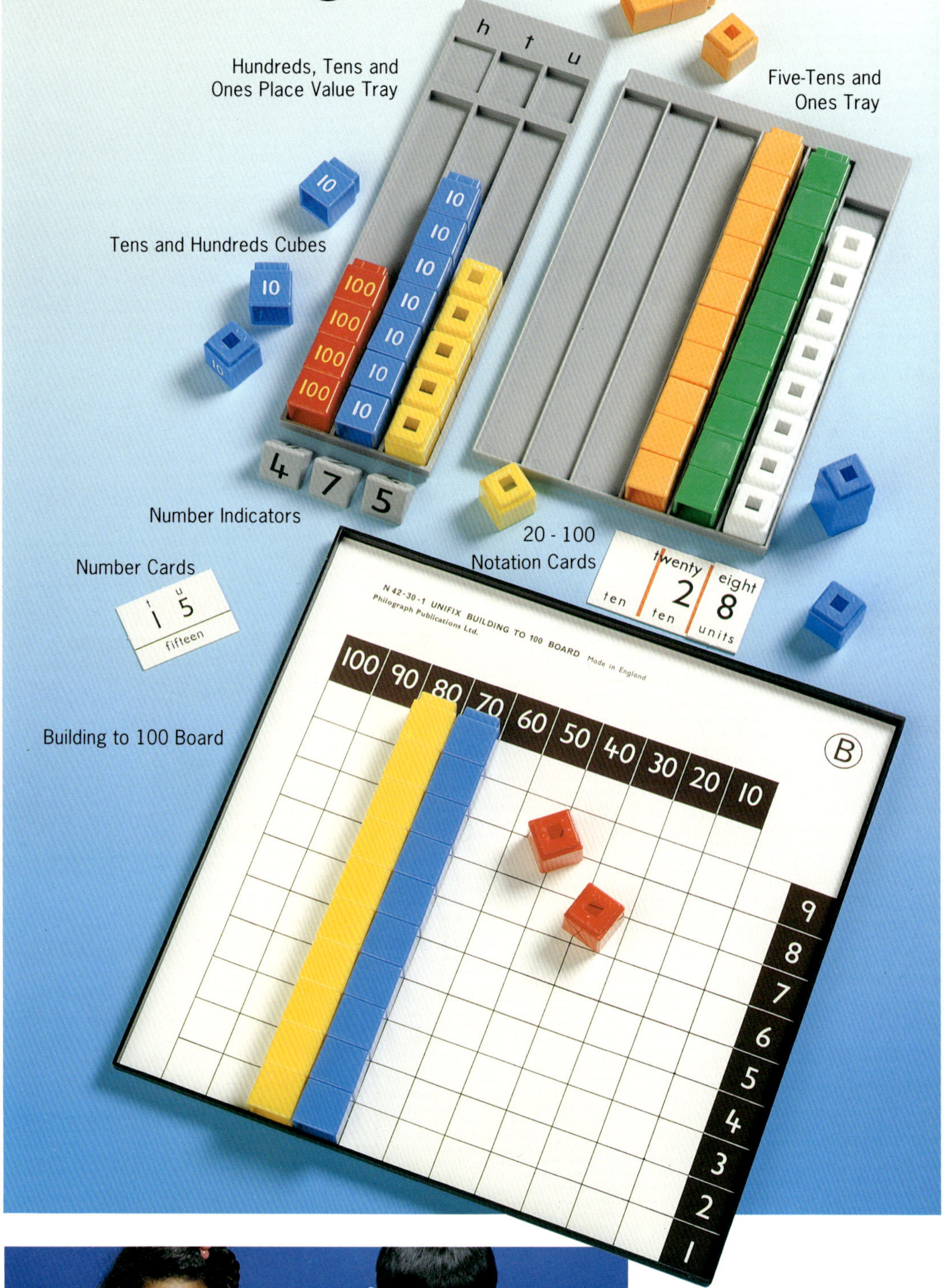

Hundreds, Tens and Ones Place Value Tray
h t u
Five-Tens and Ones Tray
Tens and Hundreds Cubes
10
100
Number Indicators
4 7 5
Number Cards
t u
1 5
fifteen
20 - 100 Notation Cards
twenty
eight
ten
2 8
ten
units
Building to 100 Board
N 42-30-1 UNIFIX BUILDING TO 100 BOARD
Philograph Publications Ltd.
Made in England
100 90 80 70 60 50 40 30 20 10
B
9 8 7 6 5 4 3 2 1

Tens
Ones
UNIFIX

Blackline Master #2 and #3

WILLARD

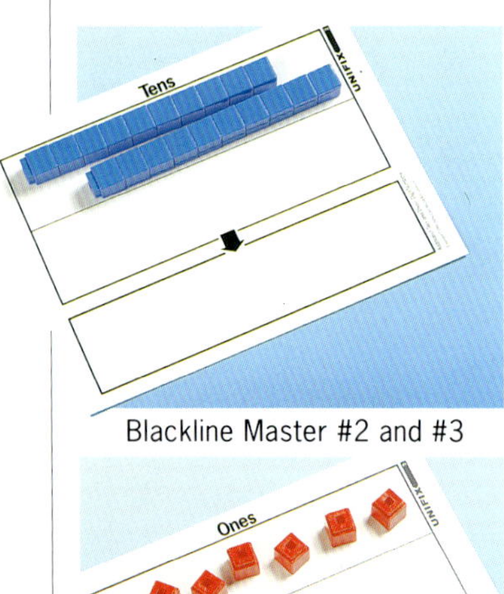

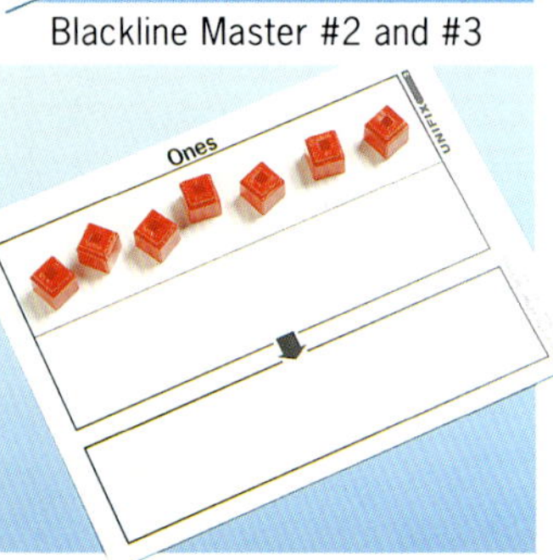

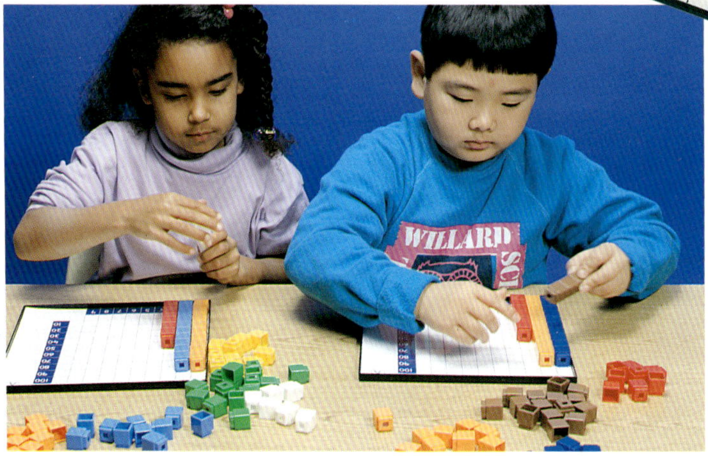

Addition of Two- and Three-Digit Numbers

Lesson 1

Objective:

Students will gain an understanding of the meaning of adding two-digit numbers without the need to trade units.

Materials:

UNIFIX CUBES

UNIFIX NUMBER CARDS

UNIFIX 20 - 100 NOTATION CARDS

UNIFIX FIVE-TENS AND ONES TRAYS

Activities:

Distribute a quantity of Unifix Cubes, Number Cards, 20 - 100 Notation Cards and a Five-Tens and Ones Tray to each student.

Write two-digit numbers such as 15 + 13 on the chalkboard.

Ask the students to show the meaning of the numbers by making two towers of ten with the Unifix Cubes and placing five cubes next to one tower of ten and three cubes next to the other tower of ten.

Then ask them to place the appropriate Number Card under the tens and ones.

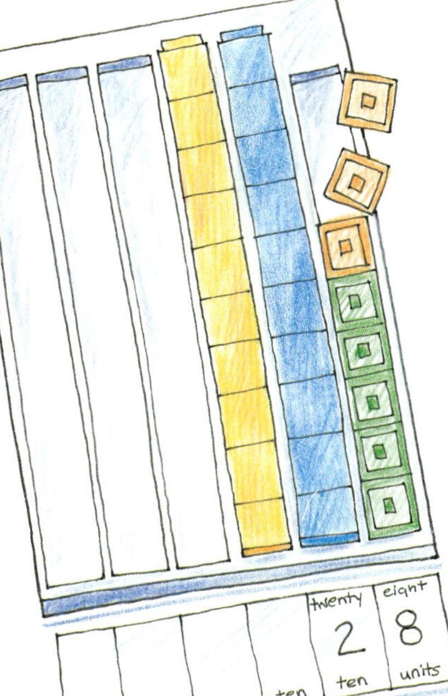

The towers of ten and ones should then be placed in the Five-Tens and Ones Tray to show the operation of addition as *"part + part = the whole."* Ask the students to place the Unifix 20 - 100 Notation Cards below the tray to show the sum of the addition problem.

Lesson 2

Objective:

Students will gain an understanding of the meaning of addition with two-digit numbers using the long form of addition that does not require the trading of ten ones for one ten, and the short form, that does require trading ten ones for one ten.

Materials:

UNIFIX CUBES

UNIFIX BUILDING TO 100 BOARDS

BLACKLINE MASTER #2 AND #3

Activities:

Prepare copies of Blackline Masters #2 and #3 for the students in your class.

Distribute a supply of *Unifix Cubes, Blackline Master #2 and #3* and the *Unifix Building to 100 Board.*

Write a problem such as 22 + 14 on the chalkboard.

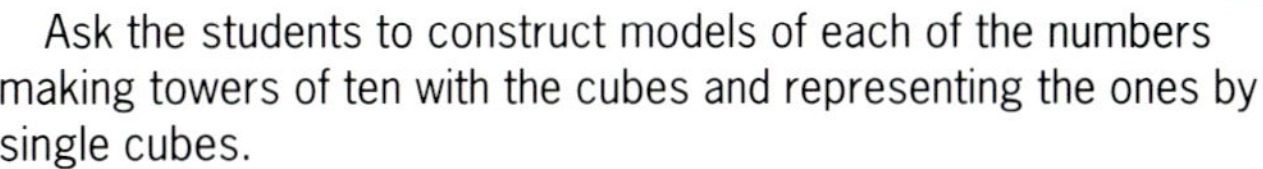

Ask the students to construct models of each of the numbers making towers of ten with the cubes and representing the ones by single cubes.

The towers of ten and the ones for each number are placed on the blackline masters with 22 above 14. Then the towers and cubes are pushed down into the bottom rectangle.

Ask the students to count the number of ones and record the number 6. Then count the number of tens and record the number 30. Lastly, the students are to record the sum. This is the long form of addition and should be demonstrated before students learn the short form of adding.

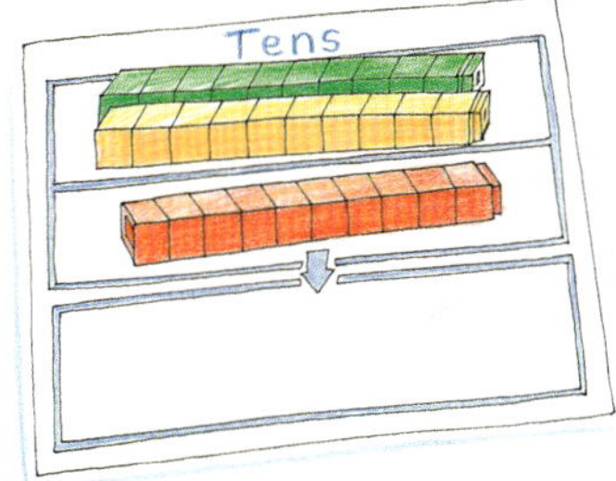

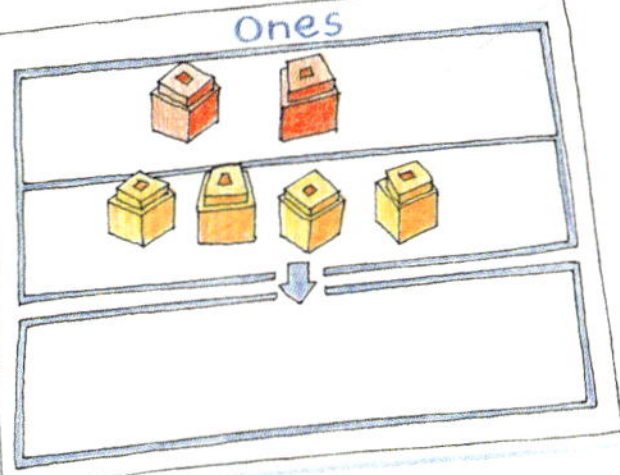

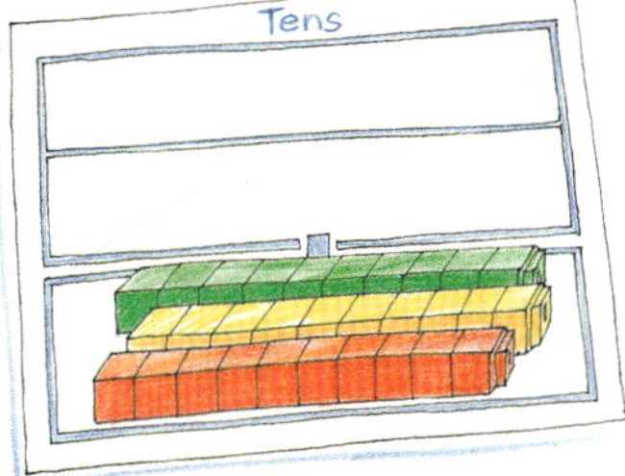

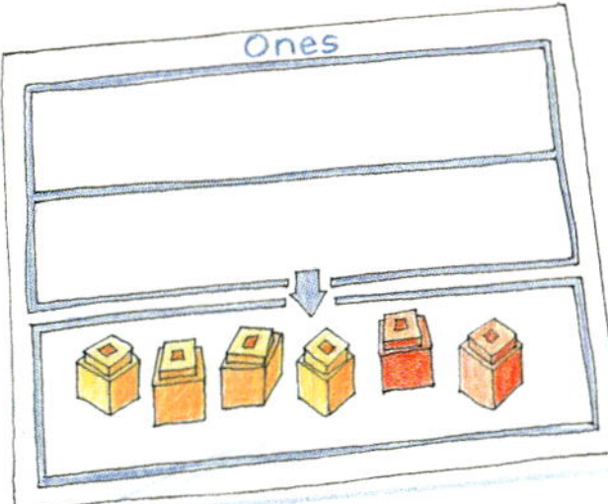

Students will learn the short form of adding by trading ten ones for one ten. Given a problem such as 26 + 17, the students model the number by making towers of tens and ones with Unifix Cubes.

The models are placed on either side of the Unifix Building to 100 Board to show the sum. First the tens and ones that show 26 are placed on the board. Next the ten and ones for 17 are placed on the board. Students will see that they have too many ones and will then trade 10 ones for one tower of ten. The new tower of ten is placed with the other tens and there remains three cubes in the ones column.

The board now shows that the sum is 43.

The student records the action by writing a one over the tens in the problem and writing the sum.

Addition of Two - and Three - Digit Numbers

Lesson 3

Objective:

Students will gain an understanding of the meaning of addition with three-digit numbers in which trading ten ones for one ten occurs in the problem.

Materials:

UNIFIX CUBES

UNIFIX TENS AND HUNDREDS CUBES

UNIFIX HUNDREDS, TENS AND ONES PLACE VALUE TRAYS

UNIFIX NUMBER INDICATORS

Activities:

Before beginning this activity, distribute a supply of *Unifix Cubes, Ten and Hundreds Cubes, Hundreds, Tens and Ones Place Value Trays* and *Number Indicators* to your students.

Students need to learn to look at three-digit numbers and see if there is a need to trade the ones for a ten.

Present the students with two problems such as: 347 + 128 or 345 +124.

$$\begin{array}{r} 347 \\ +128 \\ \hline \end{array} \quad \text{or} \quad \begin{array}{r} 345 \\ +124 \\ \hline \end{array}$$

Which of the problems would require the need to trade 10 ones for one ten?

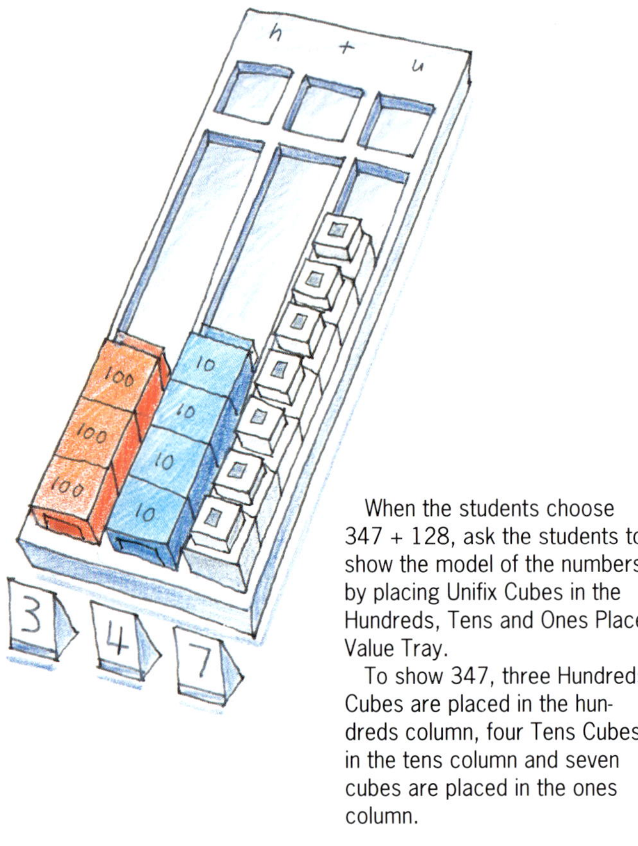

When the students choose 347 + 128, ask the students to show the model of the numbers by placing Unifix Cubes in the Hundreds, Tens and Ones Place Value Tray.

To show 347, three Hundreds Cubes are placed in the hundreds column, four Tens Cubes in the tens column and seven cubes are placed in the ones column.

Next, the students add 128 by placing one more cube in the hundreds column and two more cubes in the tens column.

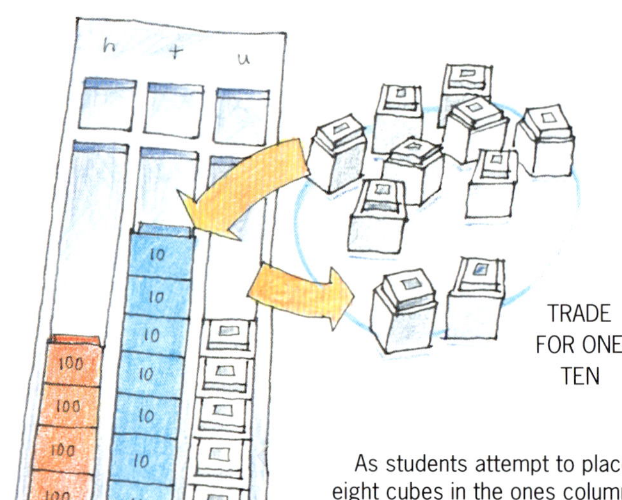

TRADE FOR ONE TEN

As students attempt to place eight cubes in the ones column, they will see that there is a need to trade ten ones for one Ten Cube. This ten is placed in the tens column and the remaining five cubes are placed in the ones column.

The Number Indicators are placed below the tray to show the sum. The students then record the action by writing the number 1 above the four in the tens column in the problem and adding the numbers.

Addition of Two - and Three - Digit Numbers

Lesson 4

Objective:

Students will gain an understanding of how to trade twice when adding three-digit numbers.

Materials:

UNIFIX CUBES

UNIFIX TENS AND HUNDREDS CUBES

UNIFIX HUNDREDS, TENS AND ONES PLACE VALUE TRAYS

UNIFIX NUMBER INDICATORS

Activities:

For this activity students will need *Unifix Cubes, Tens and Hundred Cubes, Hundreds, Tens and Ones Place Value Trays* and *Number Indicators*.

Present your students with two problems such as the ones shown here.

$$\begin{array}{r} 485 \\ +249 \\ \hline \end{array} \quad \text{or} \quad \begin{array}{r} 437 \\ +249 \\ \hline \end{array}$$

Ask the students which of the problems would require the need to trade in both the tens and ones place.

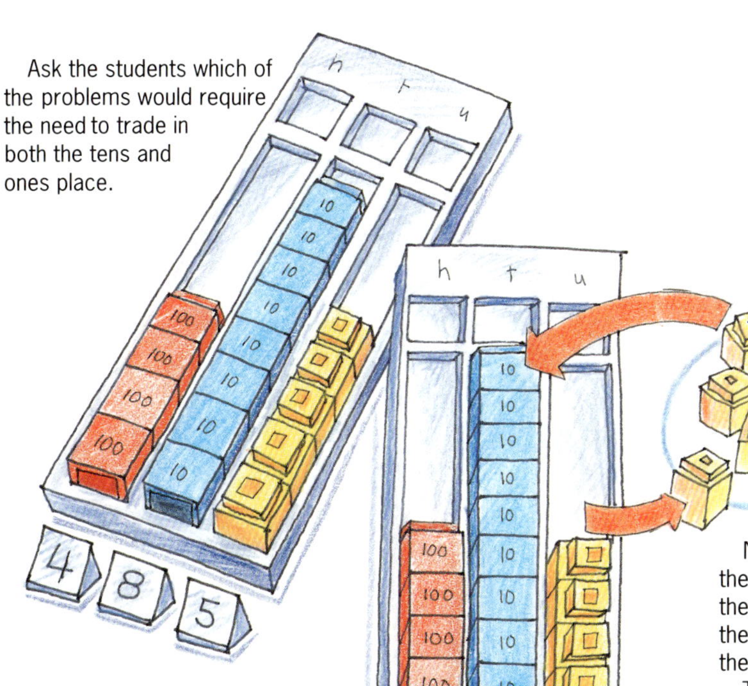

When the students pick the problem 485 + 249, ask them to place cubes in the Hundreds, Tens and Ones Place Value Tray to show the number 485.

Next, look at 249 and ask the students to place the nine ones in the one's column in the tray.

The students can see that there are too many ones for the column so they must now trade ten ones for a Ten Cube.

The Ten Cube is placed in the tray in the tens column and the four remaining cubes are placed in the ones column.

Next the four tens are placed in the tray.

The students see there is a problem because there are already nine tens in this column. Students should now see that it is necessary to trade 10 tens for one Hundred Cube.

The new Hundred Cube is placed in the hundreds column and the three remaining tens are placed in the tens column.

TRADE TEN 10'S FOR ONE HUNDRED.

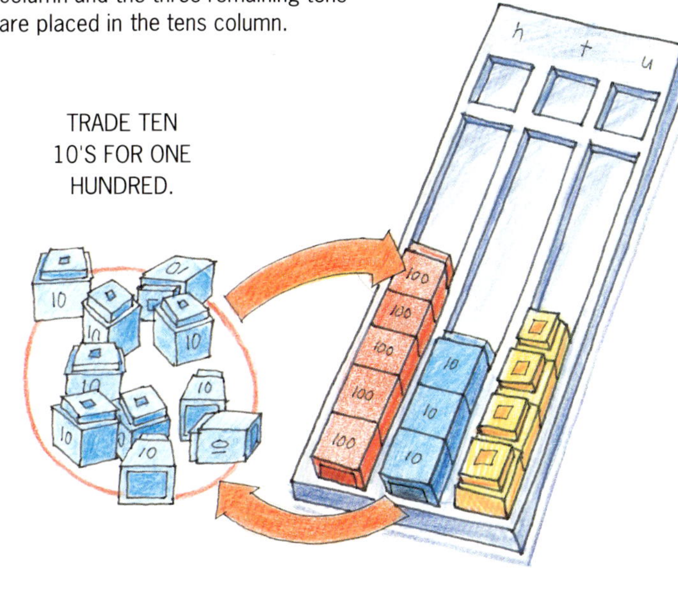

Now the students place the last two Hundreds Cubes in the hundreds column and place the Number Indicators below the tray to show the sum is 734.

The students now record the action by writing a number 1 over the tens and also a number 1 over the hundreds in the written problem and writing the sum.

Record

$$\begin{array}{r} {\scriptstyle 1\,1\,} \\ 485 \\ +249 \\ \hline 734 \end{array}$$

9 Subtraction of Two- and Three-Digit Numbers

The major trouble spot in learning to subtract numbers with two or more digits is knowing when there is a need to trade a higher unit for ten of the smaller units.

First it is important that students learn to look at the digits before they start to subtract to determine whether a trade is necessary.

Activities with manipulative materials are needed to help students learn how to trade and when to trade in subtraction.

The use of materials in teaching the understanding of subtraction can avoid learning the operation by simply following mechanical steps with little or no meaning.

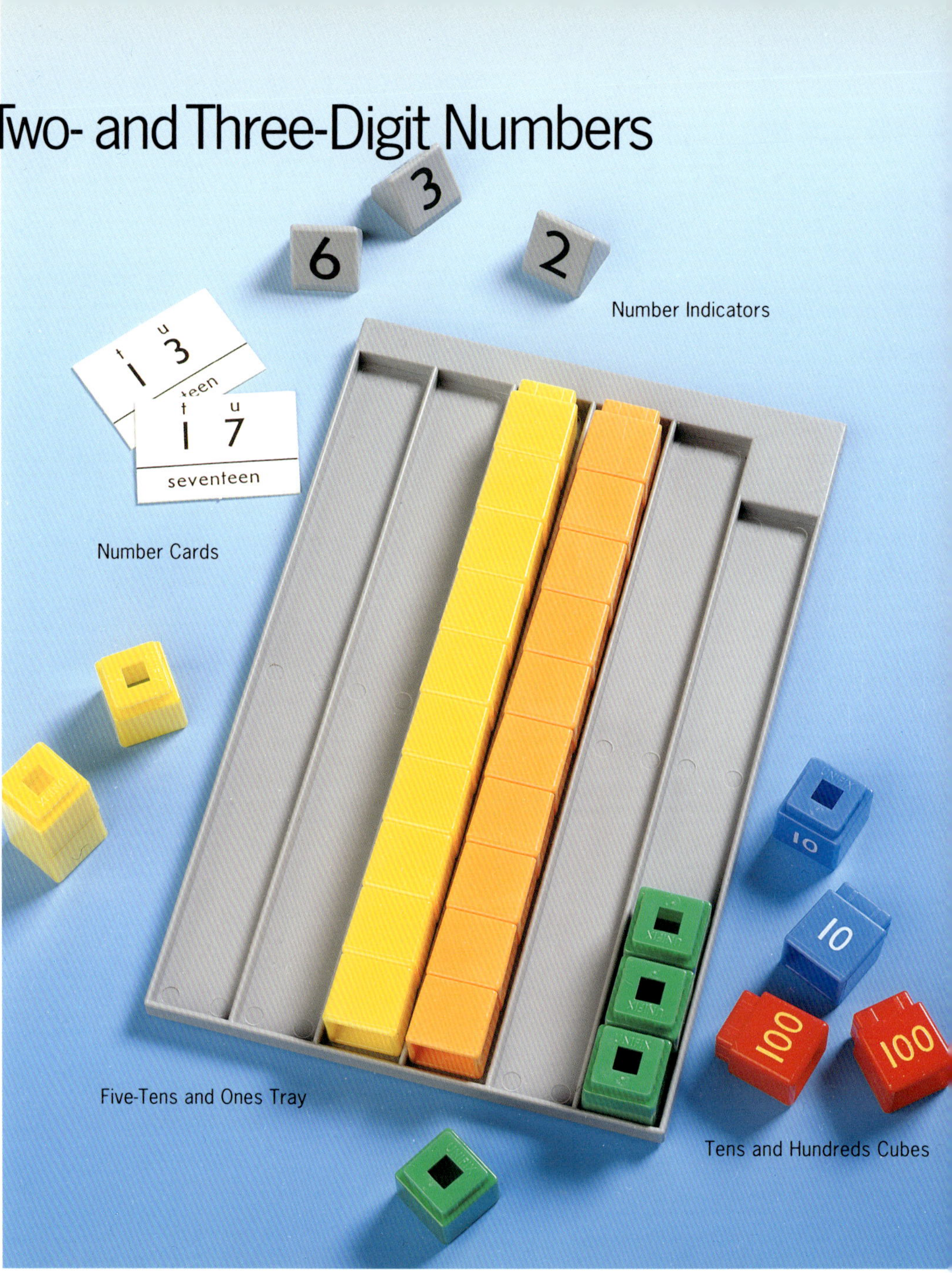

6
3
2
Number Indicators
t u
1 3
teen
t u
1 7
seventeen
Number Cards
Five-Tens and Ones Tray
10
10
100
100
Tens and Hundreds Cubes

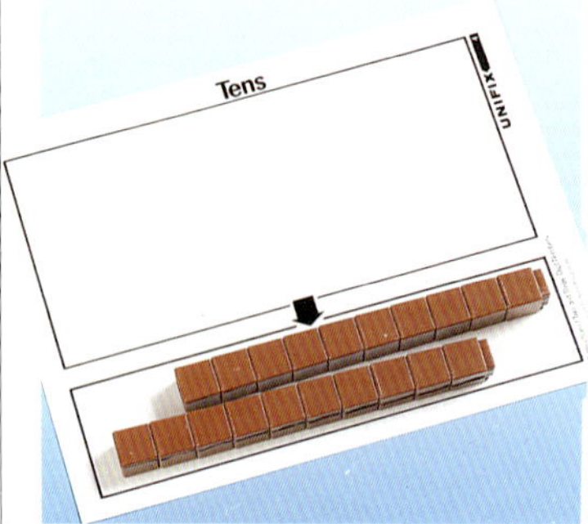

Tens
UNIFIX
Blackline Master #4

Ones
UNIFIX
Blackline Master #5

Hundreds
Tens
Ones
UNIFIX
Blackline Master #6

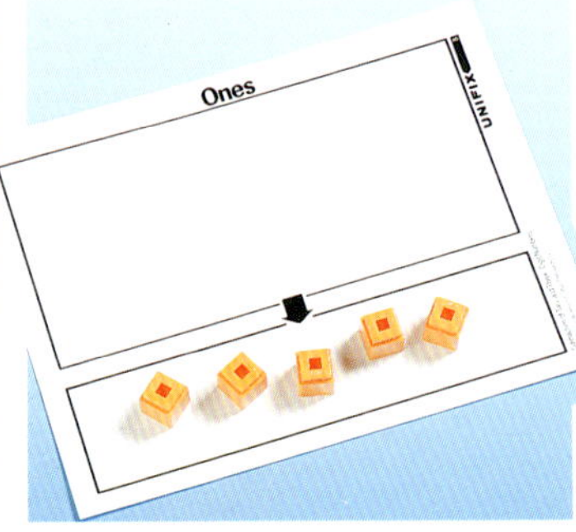

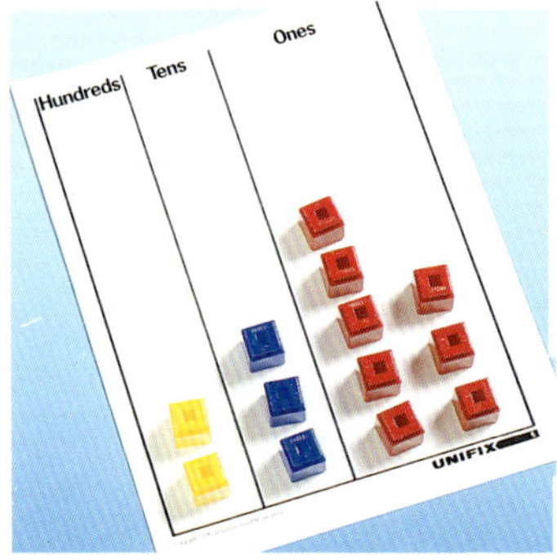

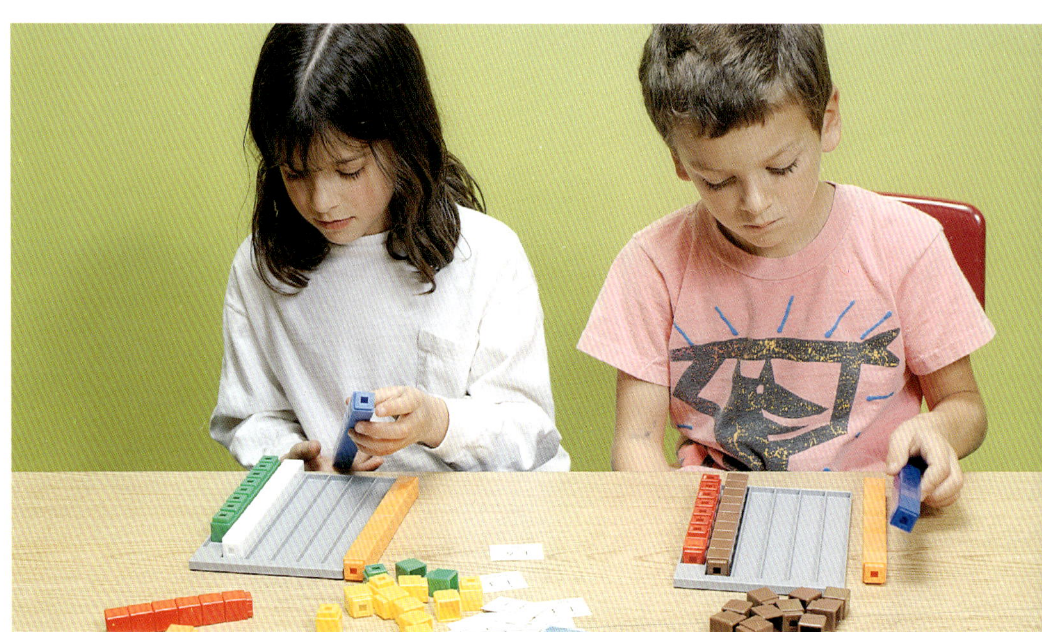

Subtraction of Two- and Three-Digit Numbers

Lesson 1

Objective:

Students will gain an understanding of subtracting two-digit numbers without the need to trade units.

Materials:

UNIFIX CUBES

UNIFIX FIVE-TENS AND ONES TRAYS

UNIFIX NUMBER CARDS

Activity

Students can work together in small groups during this activity or work individually.

Distribute a supply of *Cubes*, *Trays* and *Number Cards* to each student or group of students.

Give the students a problem such as 38 - 25 = ? and ask the students by using the cubes, trays and number cards, to model the number 38 by making three towers of ten with the Unifix Cubes and eight single cubes.

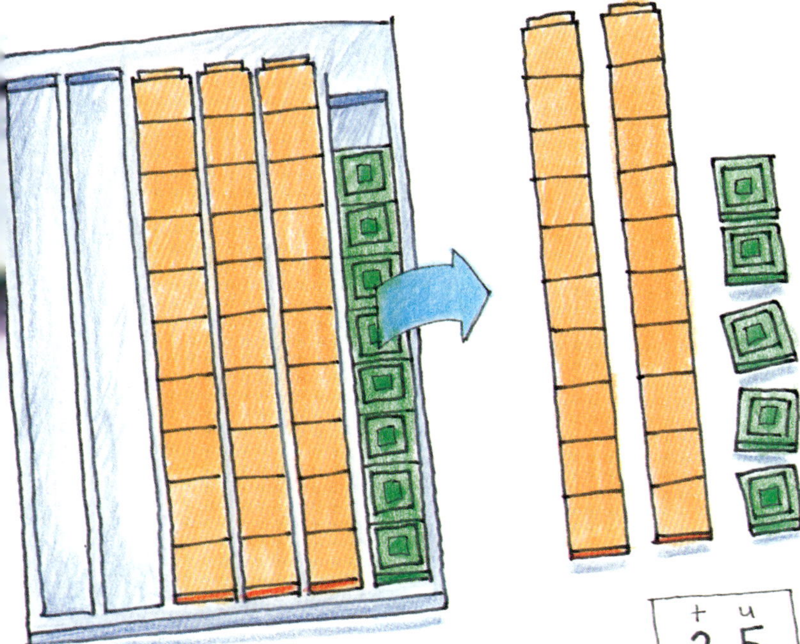

Students should place the towers and single cubes in the Five-Tens and Ones Tray then "take away" five cubes from the ones column and two tens from the tens columns and place the Number Card 25 under the tower and cubes.

Place the Number Card 13 beneath the tray.

The answer to the problem is now shown in the tray and the cubes outside of the tray show the amount that was taken away.

This type of action will help the students see the meaning of subtraction. By placing Unifix Number Cards under the tray and under the cubes that were taken away students will see the number of each of the parts.

The students should record the activity by completing the problem with pencil and paper.

Subtraction of Two - and Three - Digit Numbers

Lesson 2

Objective:

Students will gain an understanding of trading one ten for ten ones when needed in subtracting two-digit numbers.

Materials:

UNIFIX CUBES

BLACKLINE MASTERS #4 AND #5

Activities:

The first step in learning to subtract numbers is to look to see if the numbers can be subtracted without any type of trading.

0	8	3	9
-9	-2	-7	-4
N	Y	N	Y

Give your students problems similar to the ones shown and ask: *"In the set of whole numbers can you subtract the ones?"* Students should write **N** for **"no"** or **Y** for **"yes"** under each problem.

48	60	81	98
-23	-27	-37	-26
Y	N	N	Y

In the next step, give the students several problems of two-digit numbers and again ask the students if they can subtract the ones. Students respond by writing **Y** or **N** under each problem.

Duplicate enough copies of *Blackline Masters #4 and #5* for your students.

In this activity, students will show the work involved in subtracting two-digit numbers that require trading one ten for ten ones.

Give the students a problem such as 34 - 18 = ? and ask the students to show a model of 34 by making three towers or trains of ten and four single cubes.

Ask the students to place the two blackline masters side by side and place the towers or trains on the Blackline Master marked "tens" and the four single cubes on the blackline master marked "ones."

To begin the subtraction of 18 from 34, students should note that they are not able to take away eight ones when there are only four ones.

A tower or train of ten from the Tens Blackline Master must be taken apart and the ten cubes placed on the Ones Blackline Master. The students can now take away the eight ones and the one ten.

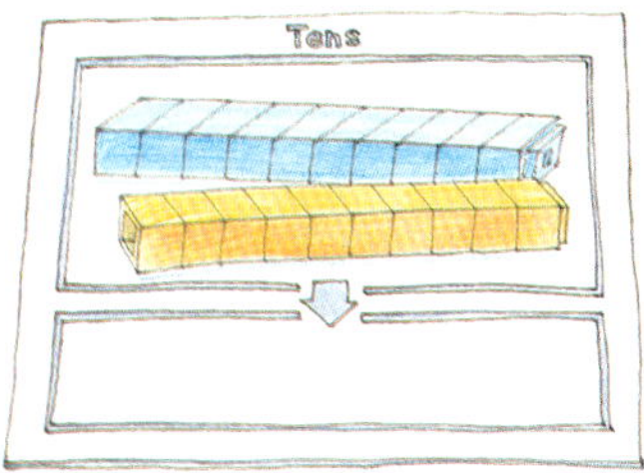

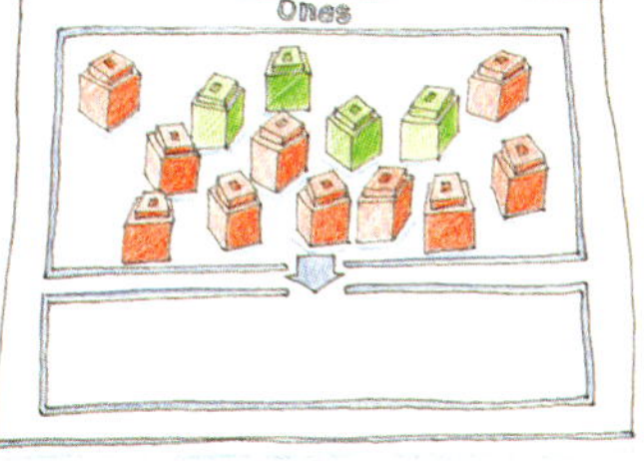

When the tower of ten and eight cubes are pushed down to the bottom sections, the answer of 16 is shown on the top part.

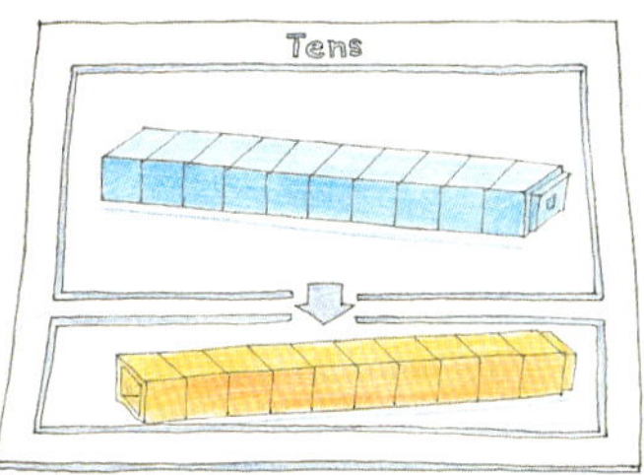

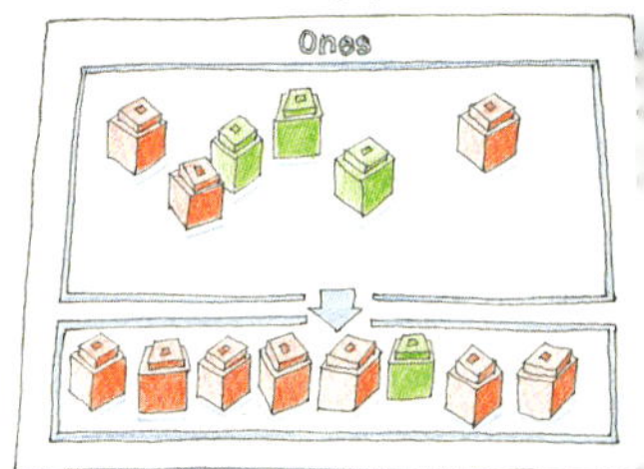

The students should then record the activity on paper by crossing out the 3 and writing a 2 and then writing a 1 by the 4.

Record

2 ~~3~~ 14
-18
16

Lesson 3

Objective:

Students will gain an understanding of subtracting three-digit numbers when there is a need to trade a ten for ten ones.

Materials:

UNIFIX CUBES

UNIFIX TENS AND HUNDREDS CUBES

UNIFIX BLACKLINE MASTER #6

UNIFIX NUMBER INDICATORS (OPTIONAL)

Activities:

Prepare copies of *Blackline Master #6* for the students in your class.

To help students learn to look at the problem before they start to subtract, give the students several problems and ask them to decide which problems require a trade of one ten for ten ones.

Ask the students: *"Can you subtract the ones?"*

Then tell the students to write **N** for **"no"** or **Y** for **"yes"** under each problem.

329	581	608	386
-145	-237	-294	-149
Y	N	Y	N

For this activity distribute *Unifix Cubes, Unifix Tens and Hundreds Cubes*, copies of *Blackline Master #6* and *Number Indicators* to the students.

Write a problem such as 386 - 149 = ? on the chalkboard.

Ask the students to show the number 386 on the blackline master by placing three Hundreds Cubes in the hundreds column, eight Tens Cubes in the tens column and six single Unifix Cubes in the ones column.

Call attention to the problem on the chalkboard, noting that 149 must be subtracted from the cubes on their blackline master.

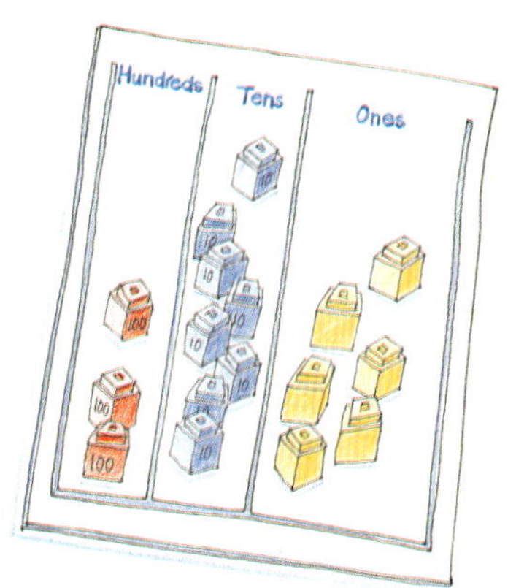

The students will note that they cannot take away nine cubes from the six cubes in the ones column.

Therefore, a Tens Cube must be traded for 10 single Unifix Cubes, and placed in the ones column with the other six single cubes.

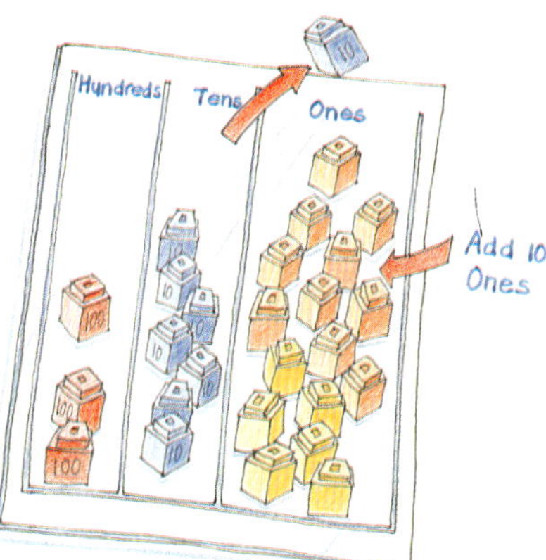

After the trade, the students can then complete the subtraction by removing 9 ones, 4 tens and 1 hundred.

Students then determine how many hundreds, tens and ones remain on the blackline master and place Number Indicators below the columns to show the answer.

Students should then record the activity they have just completed by writing the problem and then crossing out the 8 and writing in 7 and writing a 1 next to the 6.

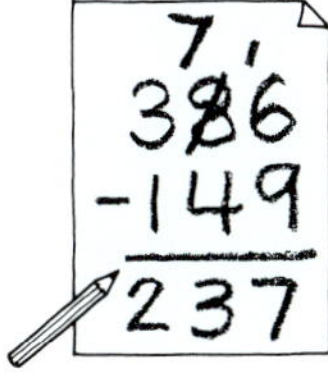

Lesson 4

Objective:

Students will gain an understanding of subtracting when zero is one of the digits.

Materials:

UNIFIX CUBES
UNIFIX TENS AND HUNDREDS CUBES
UNIFIX BLACKLINE MASTER #6
UNIFIX NUMBER INDICATORS

Activities:

Write problems similar to the ones shown below on the chalkboard. Ask the students to indicate in which place they would have to trade to solve each problem. It is important to demonstrate problems containing zeros since they cause the most difficulty in subtraction.

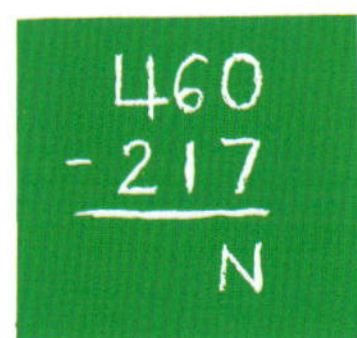

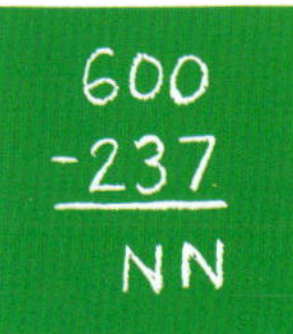

Distribute copies of *Unifix Blackline Master #6*, a supply of *Unifix Tens and Hundreds Cubes, Unifix Cubes* and *Unifix Number Indicators* to the students. Now write a problem such as 308 - 145 = ? on the chalkboard and ask the students to model the number 308 on Unifix Blackline Master #6.

308
-145

The students should place three Hundreds Cubes in the hundreds column and eight Unifix Cubes in the ones column.

Students start the subtraction process by taking away five cubes in the ones column. However, they cannot take away four tens since there are no tens in the tens column.

At this point, students should see that it is necessary to trade one Hundreds Cube for ten Tens Cubes.

Having placed the ten Tens Cubes in the tens column, students complete the exercise by taking away four tens and one hundred.

The blackline master now shows the answer and Number Indicators can be placed below each column to show the number.

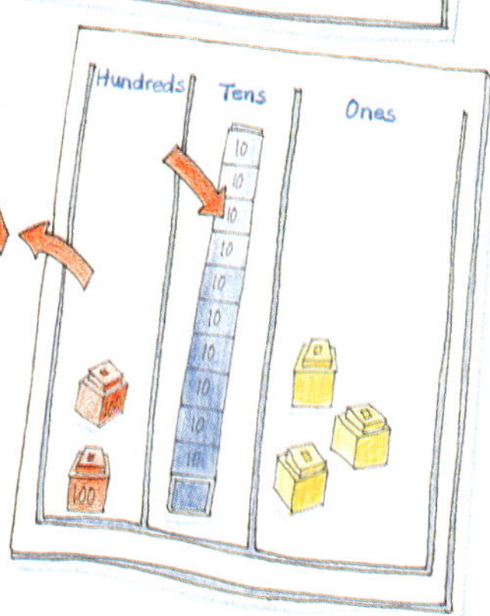

The students should then record this activity and cross out the 3, write 2 above the 3 and then write 1 next to the zero.

10 Meaning of Multiplication

The operation of multiplication is another way to add when the addends are all the same number. Activities with manipulative materials help students learn the meaning of multiplication. As students learn the meaning of multiplication, they also learn that multiplication is commutative which means you can switch the factors and the product will remain the same.

Note:

Mathematics manipulatives can create bridges or connectors from the concrete world of real things to the abstract, mathematical symbolic world. Connectors are all materials, games and activities that teachers use to help students "see" into and become competent in the world of mathematics.

See Chapter 16, page 57 for a graphic representation of the concept of connectors.

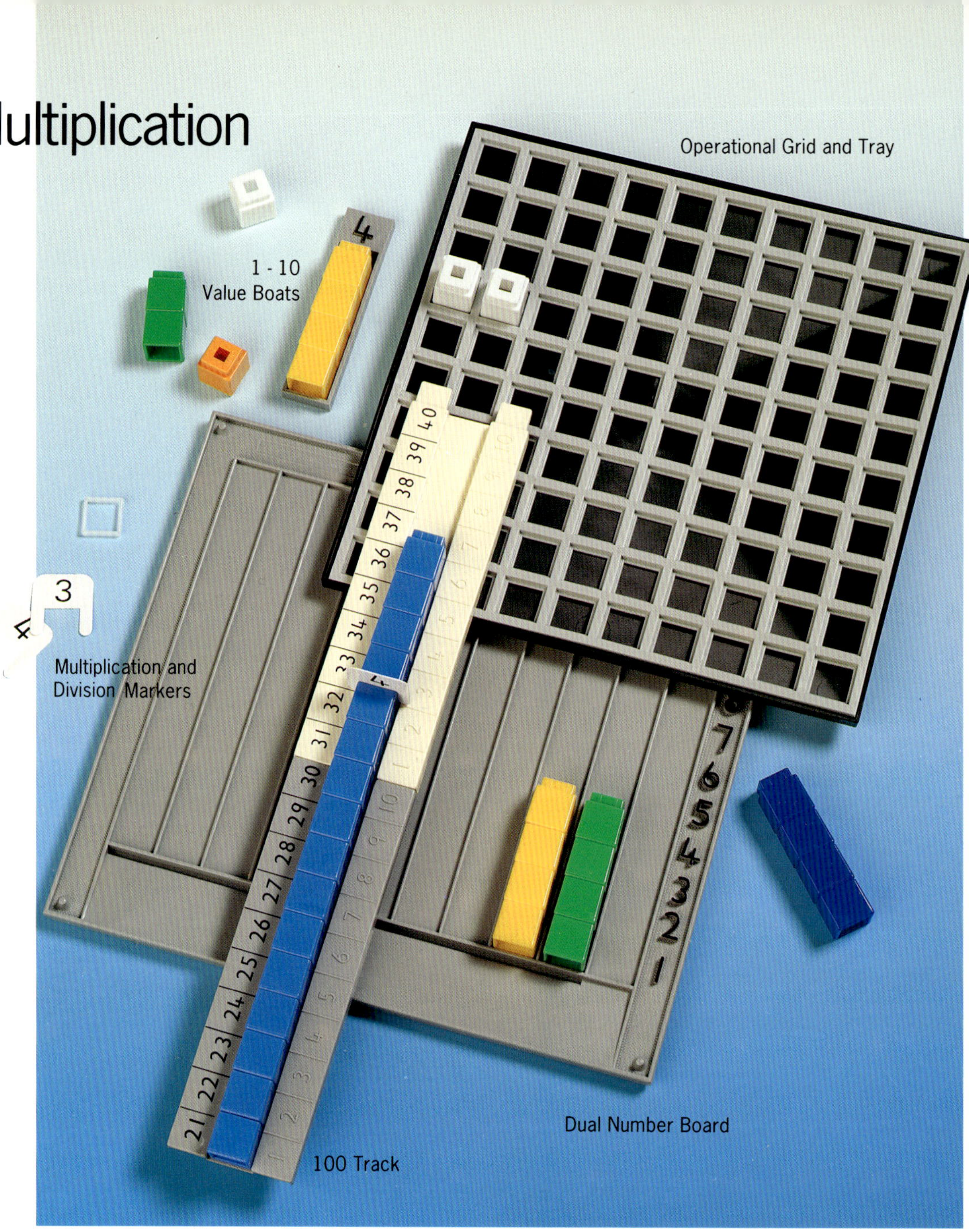

Operational Grid and Tray

1 - 10 Value Boats

Multiplication and Division Markers

Dual Number Board

100 Track

Meaning of Multiplication

Lesson 1

Objective:

Students will gain an understanding of how equal addends can be written as multiplication facts.

Materials:

UNIFIX CUBES

UNIFIX DUAL NUMBER BOARDS

UNIFIX 1-10 VALUE BOATS

Activities:

Prepare student worksheets with frame arithmetic problems in which all of the frames are the same. Instruct the students to find the number that goes in the frames to make each sentence true.

Example: □ + □ + □ = 12

Student's Answer: 4 + 4 + 4 = 12

For each frame problem, ask the students to represent the sentence using *Unifix Cubes* and the *Dual Number Board.*

For example, for the sentence, 4 + 4 + 4 = 12, ask the students to make three towers of four Unifix Cubes and place the towers in the Dual Number Board, and write below on a piece of paper, 3 x 4. Then, the three towers are joined to form one tower of ten and two ones in the Dual Number Board.

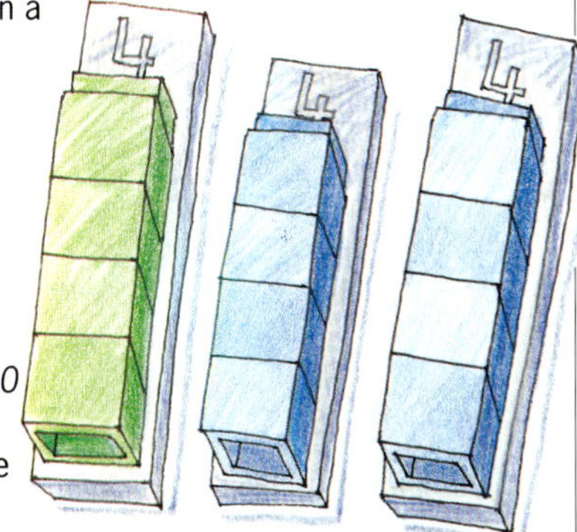

In the next activity, use the *1 - 10 Value Boats* to reinforce the concept of how equal addends can be written as multiplication facts.

Students can work in teams of two or in small groups. Using three number 4 Value Boats, fill the boats with cubes and illustrate this on the chalkboard as 4 + 4 + 4 = 12.

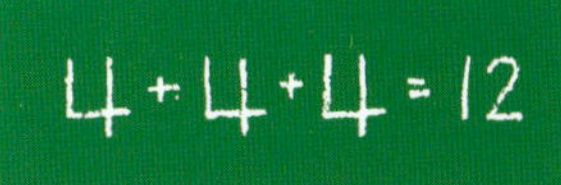

Next show the class a "faster" way to add the three fours by demonstrating the multiplication way.

If there are sufficient quantities of Value Boats available, distribute them to the students along with a quantity of Unifix Cubes and guide the students in solving other multiplication problems using the Value Boats.

If the boats are not available, ask the students to make equal towers and place Number Indicators on the tops of the towers.

Lesson 2

Objective:

Students will understand that if multiplication facts are switched, the product remains the same.

Materials:

UNIFIX CUBES

UNIFIX OPERATIONAL GRIDS

Activities:

Distribute a quantity of *Unifix Cubes* and *Operational Grids* to each student or group of students.

Write a multiplication fact such as 3 x 4 on the chalkboard and ask the students to make a rectangular model with Unifix Cubes using three horizontal rows of four cubes.

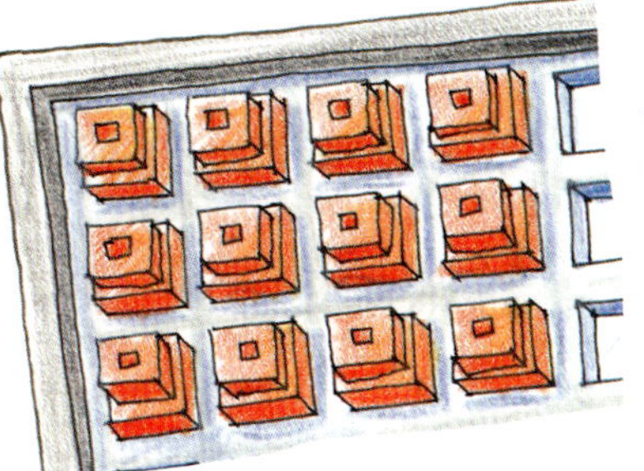

Ask the students to write the fact, 3 x 4 = 12, then turn their model into a vertical position, and write the fact, 4 x 3 = 12.

Repeat the activity for other facts.

Lesson 3

Objective:

Students will find many multiplication problems from a given number.

Materials:

UNIFIX CUBES

UNIFIX OPERATIONAL GRIDS AND TRAYS

UNIFIX 100 TRACKS

MULTIPLICATION AND DIVISION MARKERS

Activities:

Distribute 24 *Unifix Cubes* and an *Operational Grid and Tray* to each student or group of students.

Write the equation 4 x 6=24 on the chalkboard and then demonstrate how to form a rectangular array of *Unifix Cubes* using the *Operational Grid and Tray* to illustrate the equation 4 x 6 = 24.

Discuss the array: *"How many cubes across? How many rows of six are there?"*

Now ask the students to make rectangular arrays with their cubes to form different models of multiplication facts to equal 24.

Distribute *Unifix 100 Tracks, Multiplication Markers* and Unifix Cubes to the students.

Working in small groups, ask the students to put together the Unifix 100 Track up to 50. Ask the students to put Unifix Cubes on the track up to the number 36, then, using the Unifix Multiplication Markers, find as many ways they can to make 36.

To start, use the six markers and find how many sixes make 36.

Write the fact, 6 x 6 = 36 on the chalkboard. Then students can use the four markers, three markers, nine markers, etc.

11 Multiplication Facts

UNIFIX

It is obvious that students have to learn the multiplication facts before they begin multiplying with large numbers. Therefore, it is helpful for students first of all to be involved with materials that model the facts then recording the activity by writing the whole fact.

To understand and memorize these facts, students need a great deal of practice in writing the whole facts.

The following activities, used after engaging in the activities in the previous chapter, will help your students with their understanding of multiplication and the memorization of the multiplication facts.

Cubes

Multiplication and Division Markers

100 Track

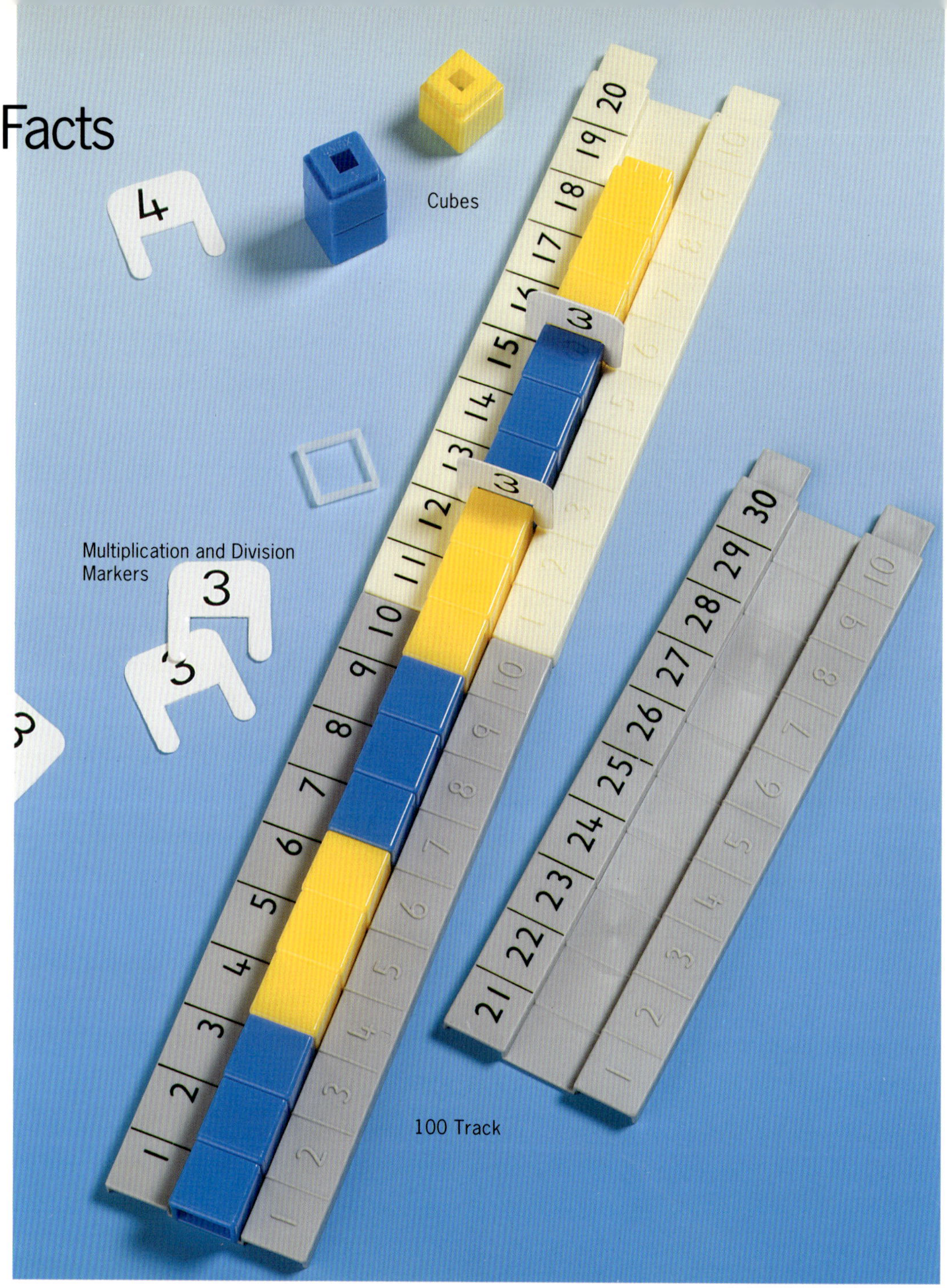

Multiplication Facts

Lesson 1

Objective:

Students will gain an understanding of and memorize all of the multiplication facts.

Materials:

UNIFIX CUBES

UNIFIX 100 TRACKS

UNIFIX MULTIPLICATION AND DIVISION MARKERS

Activities:

Distribute a supply of *Unifix Cubes, a Unifix 100 Track* and the *Number 3 Multiplication Markers* to each group of students.

Working together in groups, students put the Unifix 100 Track together. During this activity, students will model all of the multiplication facts and write the whole fact.

Ask the students to model the multiplication table of three, using two different colors of Unifix Cubes by making a train of three cubes of one color and join three cubes of another color and continuing this pattern until the train contains 27 joined cubes.

The train of 27 cubes is placed on the Unifix 100 Track.

The three's multiplication markers should be placed after every three cubes. Using this model, ask the students to write all of the multiplication facts that have three as a factor.

They should write, 1 x 3 = 3, 2 x 3 = 6, 3 x 3 = 9, etc. The activity should be continued using twos, fours, fives, sixes,sevens, eights and nines.

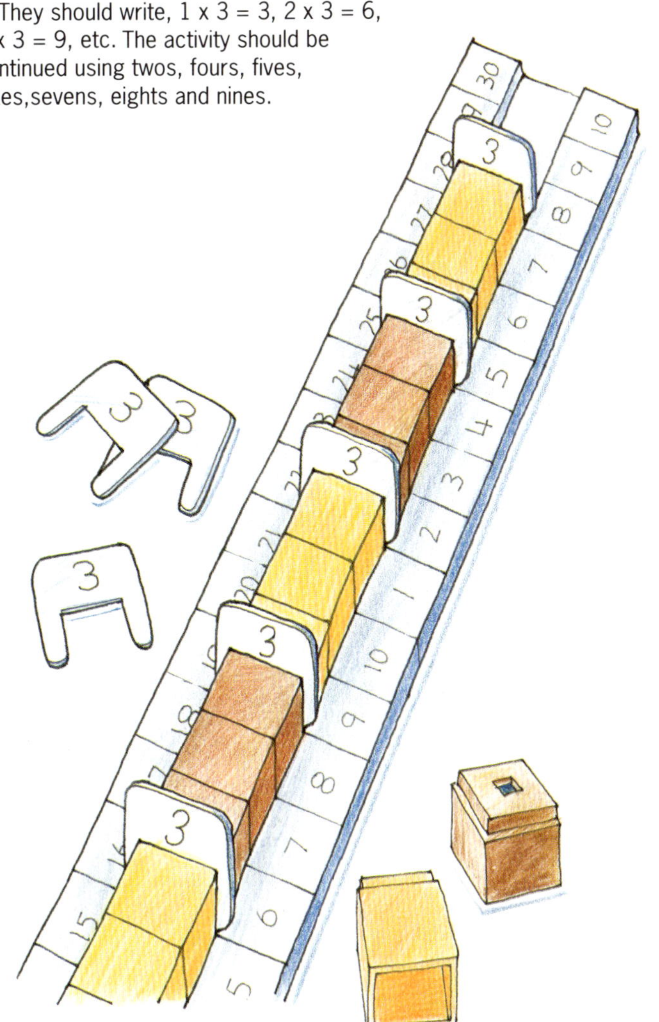

Lesson 2

Objective:

Students will engage in activities that will reinforce the concept that if the factors are reversed, the product remains the same.

Materials:

UNIFIX CUBES

UNIFIX 100 TRACKS

UNIFIX MULTIPLICATION AND DIVISION MARKERS

Activities:

Distribute a supply of *Unifix Cubes, a Unifix 100 Track* and *Multiplication Markers* to each group of students.

After assembling the 100 Track, ask the students to join together a train of 24 cubes.

Direct the students to place the train on the 100 Track and, using the number 6 Multiplication Markers, divide the train after every six cubes. Students should record this activity as 6 x 4 = 24.

After removing the number 6 markers, students should place the number 4 multiplication markers after every four cubes and record the activity as 4 x 6 = 24.

Repeat this activity with 24 cubes using the 3 markers and the 8 markers.

This basic activity should be repeated with a train of 20 cubes with four and five markers. Each time the students should write the two facts with the factors reversed.

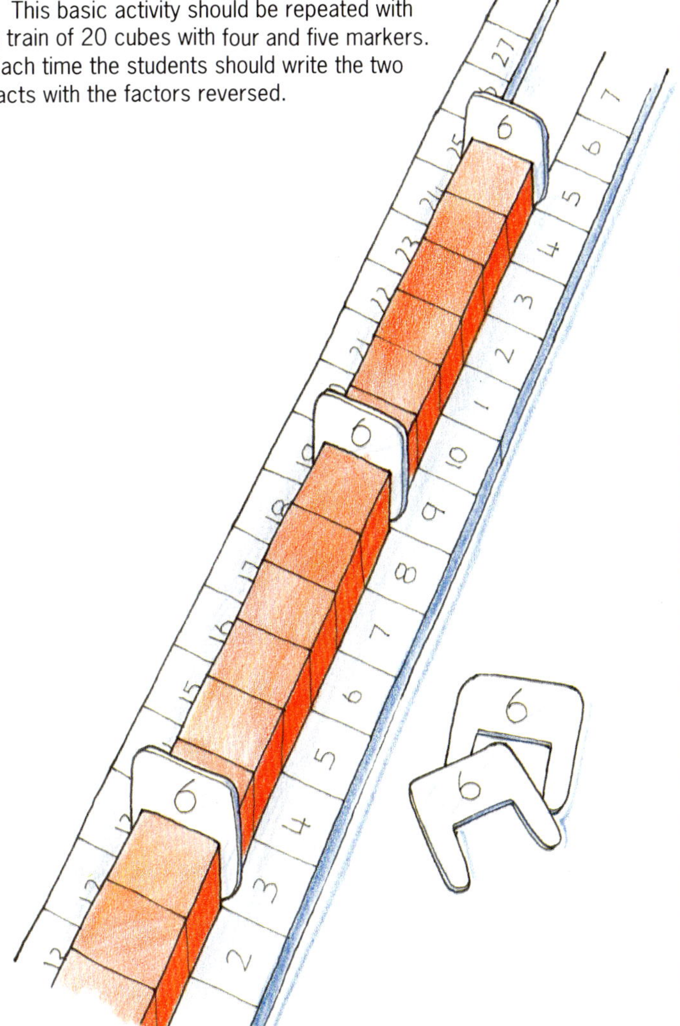

UNIFIX

12 Multiplication of One-Digit Times Two-Digit Numbers

The prerequisite for multiplying large numbers is a basic knowledge of the meaning of multiplication and the multiplication facts.

Chapters 10 and 11 offered activities that helped students understand the meaning of multiplication and to memorize the facts.

In this chapter, students will multiply tens and ones by a one-digit number through the use of Unifix manipulative materials.

One-Ten and Ones Tray

Cubes

Five-Tens and Ones Tray

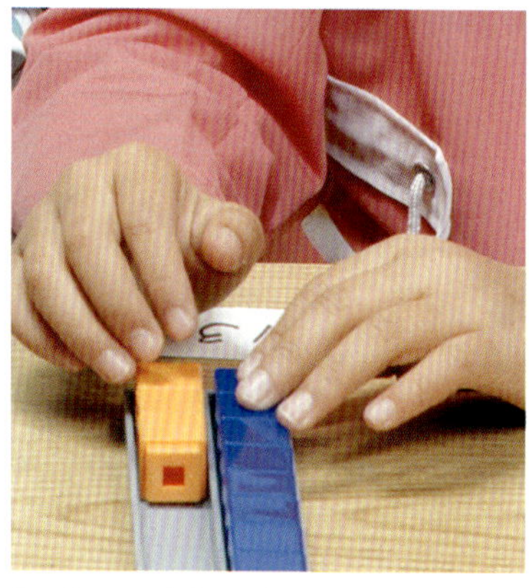

One-Ten and Ones Tray

Multiplication of One-Digit Times Two-Digit Numbers

Lesson 1

Objective:

Students will gain an understanding of multiplying one-digit times two-digit numbers with no carrying.

Materials:

UNIFIX CUBES

UNIFIX ONE-TEN AND ONES TRAYS

UNIFIX FIVE-TENS AND ONES TRAYS

Activities:

Write a problem such as 3 x 13 = ? on the chalkboard.

Working in groups, ask the students to model the number 13 in three *Unifix One-Ten and Ones Trays* with a tower of ten *Unifix Cubes* and three single cubes in each tray.

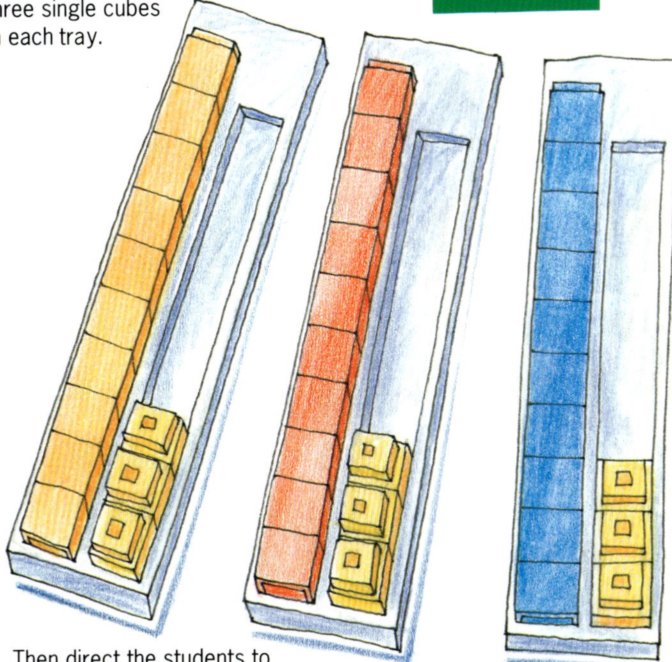

Then direct the students to write the number 13 on three pieces of paper and place them below each tray.

The towers of ten are taken out of each tray and all of the single cubes are taken out and placed in the *Five-Tens and Ones Tray*.

The model now shows 39, the product of 3 x 13. The students record their work:

Lesson 2

Objective:

Students will gain an understanding of multiplying one-digit by two-digit numbers with carrying.

Materials:

UNIFIX CUBES

UNIFIX ONE-TEN AND ONES TRAYS

UNIFIX FIVE-TENS AND ONES TRAYS

Activities:

Distribute a quantity of *Unifix Cubes* and *One-Ten and Ones Trays* to each group of students.

Write a problem such as 3 x 16 = ? on the chalkboard.

Working in groups, ask the students to model 16 in three Unifix One-Ten and Ones Trays, with a tower of ten Unifix Cubes and six single cubes in each tray.

Students should write the number 16 on three pieces of paper and place one below each tray.

The towers of ten and single cubes are then taken out of the trays. The single cubes should be joined together to form a tower of 18 cubes.

Ask the students what they think the next step in the process should be.

Point out that it is necessary to make a trade to arrive at an answer.

The students should see that it is necessary to trade ten of the 18 cubes for a tower of ten which would leave eight single cubes.

Students place the towers and cubes in the Five-Tens and Ones Tray.

The model of four towers of tens and eight single cubes now shows the product of 48.

The students record the activity as

...and should indicate that a ten was carried to the tens column by putting a number 1 above the 1 in 16.

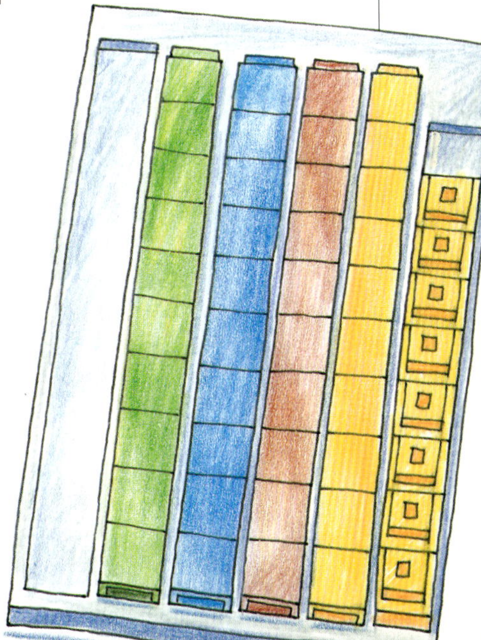

13 Meaning of Division

UNIFIX

The operation of division is the inverse or opposite operation of multiplication. However, this definition has little meaning for young students who are learning about applying division in mathematics.

In division you are given a whole set and the numbers in each subset and you must find how many subsets there are, or you are given a whole number and the number of subsets in that whole number and you must find the number in those subsets.

In the following activities, the students model the division process with Unifix manipulative materials to come to an understanding of the meaning of division.

Two concepts that help give meaning to the operation of division are the ideas of fair shares and repeated subtraction.

These concepts are presented in Lesson 1.

Note:

Mathematics manipulatives can create bridges or connectors from the concrete world of real things to the abstract, mathematical symbolic world. Connectors are all materials, games and activities that teachers use to help students "see" into and become competent in the world of mathematics.

See Chapter 16, page 57, for a graphic representation of the concept of connectors.

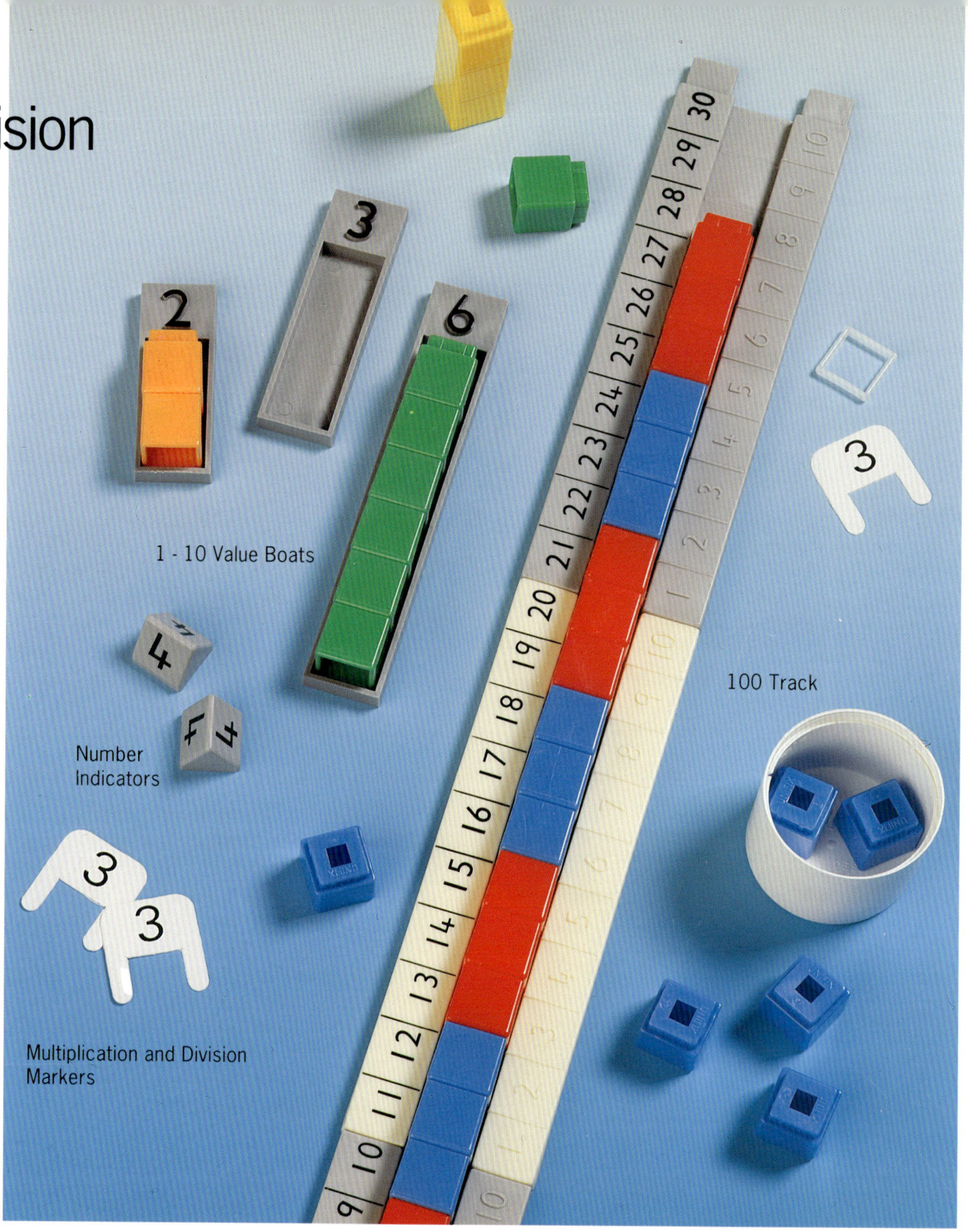

1 - 10 Value Boats

100 Track

Number Indicators

Multiplication and Division Markers

Meaning of Division

Lesson 1

Objective:

Students will achieve a better understanding of division using the fair share and repeated subtraction activities.

Materials:

UNIFIX CUBES

SMALL CONTAINERS

UNIFIX 1-10 VALUE BOATS

Activities:

Arrange the students in groups of three's.

Distribute a quantity of *Unifix Cubes* and three *small containers* to each group.

Ask the students to count out nine cubes and share them equally by placing the same amount in each of the containers.

"How many cubes will each student have in his or her container?"

It is important at this point to discuss with your students what they have just done and to record the activity.

Using the same grouping of three's, distribute 20 Unifix Cubes and the *1 to 10 Value Boats* to each group.

Ask the students to make a train of the 20 cubes.

Then ask, *"How many times can you fill the #2 Value Boat using the cubes from your 20-cube train?"*

Ask one student to tally the number of times the #2 Value Boat is filled.

Repeat this activity with the #3, #4, #5, #6, etc. Value Boats.

To reinforce the concept of successive subtraction, prepare a worksheet with the following information for each student, omitting the answers.

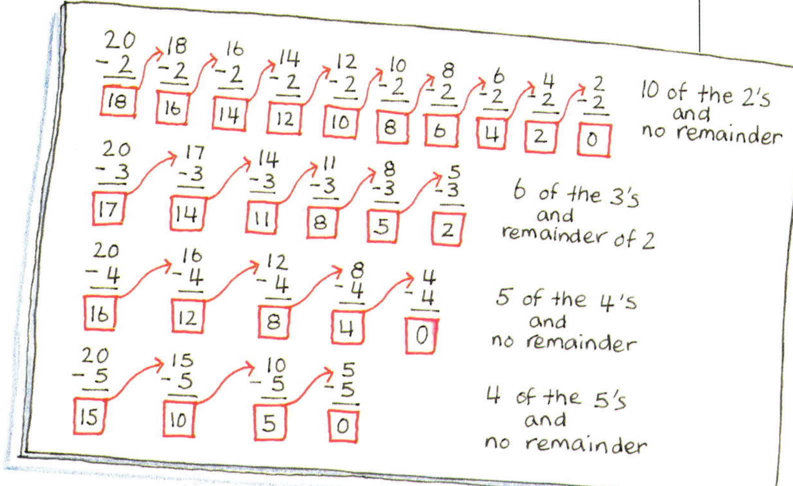

13 Meaning of Division

Lesson 2

Objective:

Students will discover the meaning of division using whole sets to a number of ten.

Materials:

UNIFIX CUBES

UNIFIX MULTIPLICATION AND DIVISION MARKERS

UNIFIX NUMBER INDICATORS

UNIFIX 100 TRACKS

Activities:

Distribute a quantity of *Unifix Cubes, Unifix Multiplication and Division Markers* and *Unifix Number Indicators* to each student or group of students.

Write the equation $8 \div 4 = ?$ on the chalkboard and tell the students that they are going to solve this problem using Unifix Cubes.

8÷4=?

Ask the student to make a tower or set of eight Unifix Cubes.

Then make subsets of four cubes from the eight cubes and place a number 4 Number Indicator on top of each subset.

"How many subsets of four do you have?"

As students see that they have two subsets, they should record the equation: $8 \div 4 = 2$

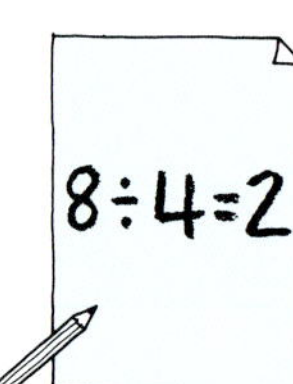

Put the equation $10 \div 2 = ?$ on the chalkboard.

10÷2=?

Next ask the students to join ten cubes and place them in the 100 Track.

Then, counting by twos, the students should place a 2 Division Marker between every second cube.

Students should then count the number of sets of two and record the activity as $10 \div 2 = 5$.

These activities can be repeated using different numbers until the students gain confidence and have an understanding of the process.

Lesson 3

Objective:

Students will gain an understanding of division with no remainder.

Materials:

UNIFIX CUBES

UNIFIX 100 TRACKS

UNIFIX MULTIPLICATION AND DIVISION MARKERS

Activities:

Put the following problems on the chalkboard:

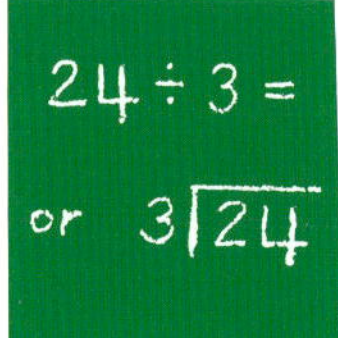

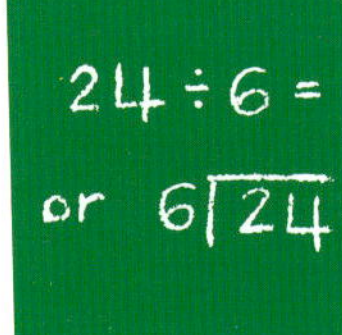

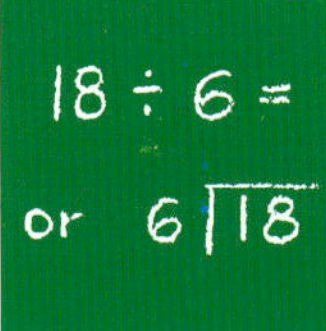

Distribute *24 Unifix Cubes, the Unifix 100 Track* and *Unifix Division Markers* to each student or group of students.

Ask the students to make a train of 24 Unifix Cubes and place the train on the Unifix 100 Track. Students should then place the 3 Unifix Division Markers between every third cube and then count the number of sets of three.

Students record

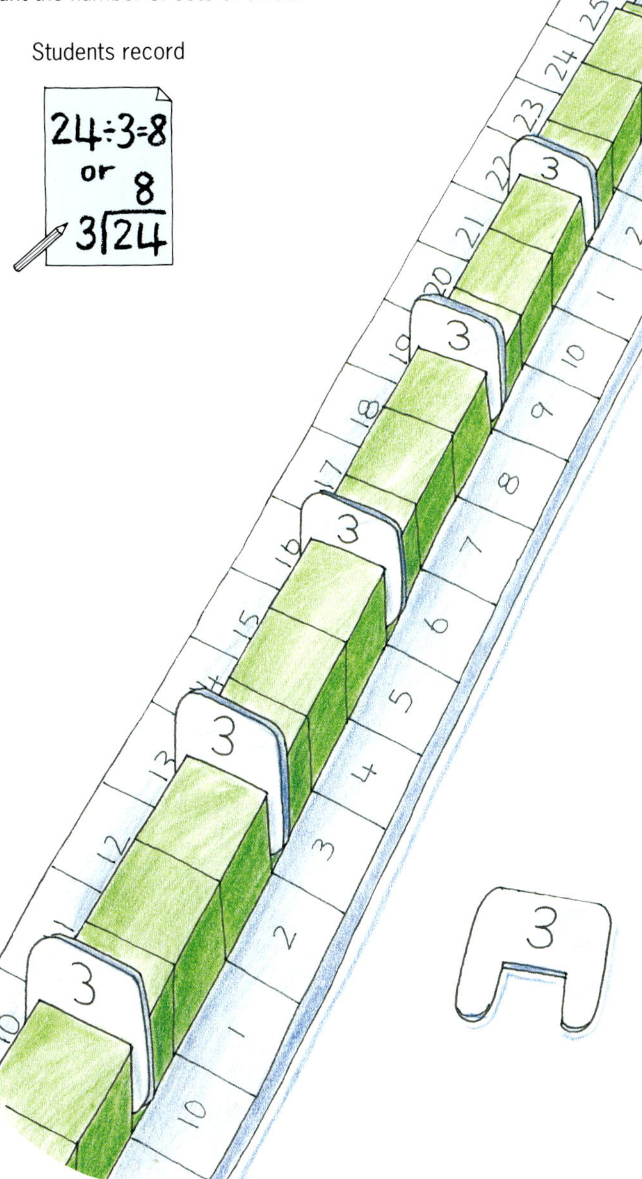

Using the same 24 cubes, the students next model 24 ÷ 6 by placing the 6 division markers after every sixth cube and then counting the number of sets of six.

The students should then record

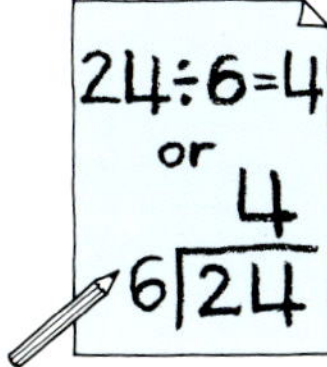

Ask the students to model the problem 18 ÷ 6 = without teacher assistance or direction.

Lesson 4

Objective:

Students gain an understanding of division with a remainder.

Materials:

UNIFIX CUBES

UNIFIX 100 TRACKS

UNIFIX MULTIPLICATION AND DIVISION MARKERS

Activities:

Write the following problem on the chalkboard:

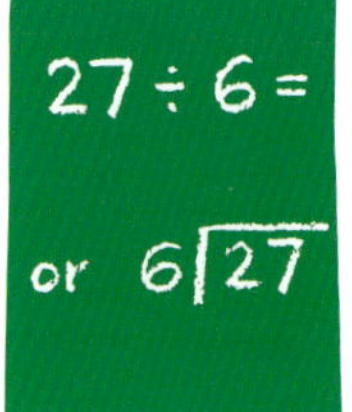

Distribute *27 Unifix Cubes, Unifix Division Markers* and *Unifix 100 Track* to each student or group of students participating in this activity.

Individually or in a group, the students make a train of 27 Unifix Cubes and place the train on the Unifix 100 Track.

Instruct the students to start at one and count six cubes and place a 6 Division Marker, count the next six cubes and place another 6 Division Marker.

Repeat this for four sets of six.

Students should see that 27 ÷ 6 is four sets of six and three cubes left over.

Illustrate the answer on the chalkboard as:

Explain to the students that the letter **"r"** stands for remainder.

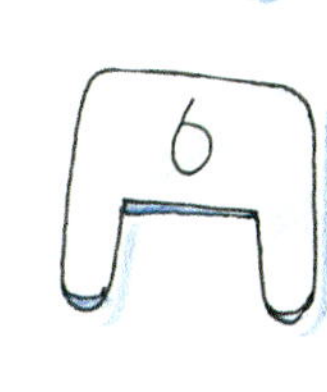

This activity should be repeated for other problems such as 37 ÷ 5, 19 ÷ 3, etc. until students understand the process of dividing with a remainder.

14 Measurement

UNIFIX

Retaining Frames

Cube Corner Units

Assembly Grids

Measurement concerns direct comparisons of objects in which decisions are made about height, weight, area, temperature, volume, capacity or time and involves the senses of touch and sight.

Students should be actively and meaningfully involved in measuring as well as writing about and discussing their conclusions in order to fully understand the concepts of measurement.

Measurement also is a problem-solving activity. We often want to know how one thing compares with another: *"Is it more? Is it less? Is it the same?"* At this point, it is important to encourage children to estimate and talk about the process of measuring.

Direct comparison between objects cannot always be made but we can use such familiar objects as Unifix Cubes, parts of the body, pencils, pieces of string and other objects as measurement tools.
It is important to provide many experiences with arbitrary or non-standard units of measurement which will lead students to discover the need for standard units.

Measurement activities can be divided into four stages:

- Making direct comparisons between objects, such as comparing the heights of two students by standing them next to each other.
- Making comparisons with non-standard units such as Unifix Cubes, crayons, pencils, hand spans, etc.
- Making comparisons of objects with standard units such as centimeters, inches, grams, ounces, yards, etc.
- Using specific or specialized units of measurement for greater accuracy and precision such as those used by scientists.

Unifix Cubes can be successfully used as a non-standard unit of measurement in determining length, perimeter, area and volume. The following activities using Unifix manipulatives will give your students practical experiences in measurement.

Lesson 1

Objective:

Students will gain an understanding of length by measuring shoes and arms with Unifix Cubes and comparing the results of their measurements.

Materials:

UNIFIX CUBES

Activities:

Have your student work together in pairs to measure each others' shoes and arms using Unifix Cubes as the non-standard unit of measurement.

Distribute 20 to 30 Unifix Cubes to each pair of students participating in this activity.

Tell the students that they are to measure the length of their shoes and their arms from elbow to the tip of the longest finger using Unifix Cubes. They should then record the number of cubes they used.

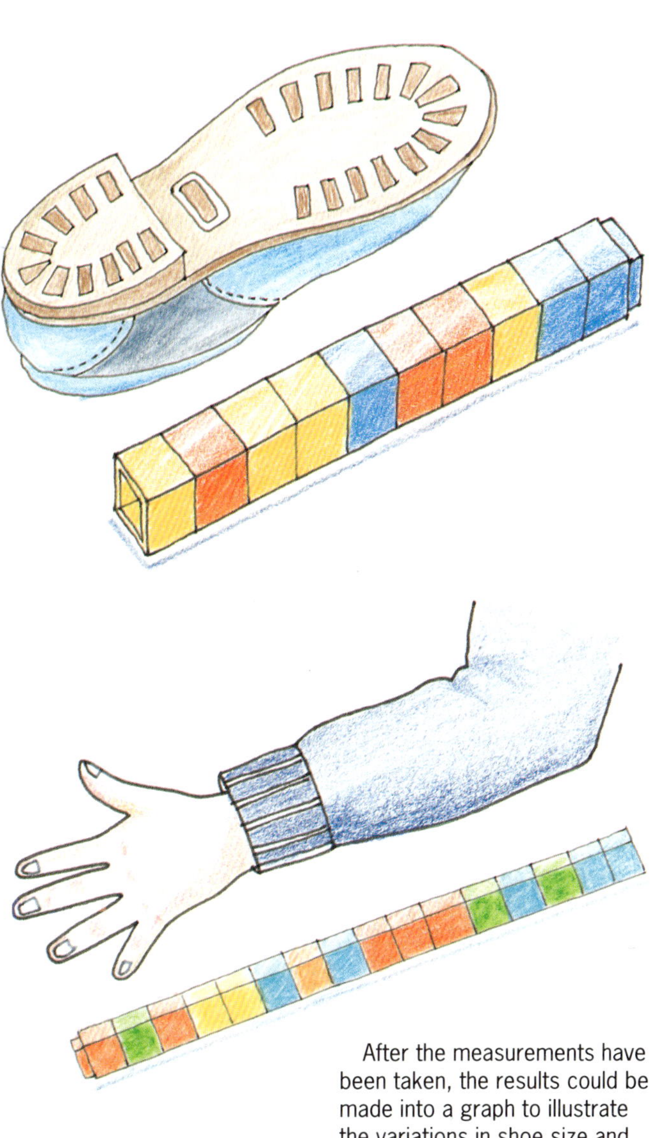

After the measurements have been taken, the results could be made into a graph to illustrate the variations in shoe size and arm length among the students participating in the activity.

Lesson 2

Objective:

Students will gain an understanding of the concept of finding the perimeter of a figure.

Materials:

UNIFIX CUBES

UNIFIX CUBE CORNER UNITS

Activities:

Distribute *12 Unifix Cubes* and a quantity of *Unifix Cube Corner Units* to each student.

If necessary, demonstrate how the Unifix Cubes and the Corner Units should be connected. Then ask the students to make as many different shapes as they can by joining 12 cubes.

After the shapes are made, ask the students to find the distance, or perimeter, around each shape.

Demonstrate this by counting around one of the figures with the students.

Explain that this is called the perimeter of the shape.

Each student should record his or her results. Then ask the students to exchange shapes and count the perimeter to verify the count.

Lead a discussion about why the perimeter changes even though the number of cubes is the same in each shape.

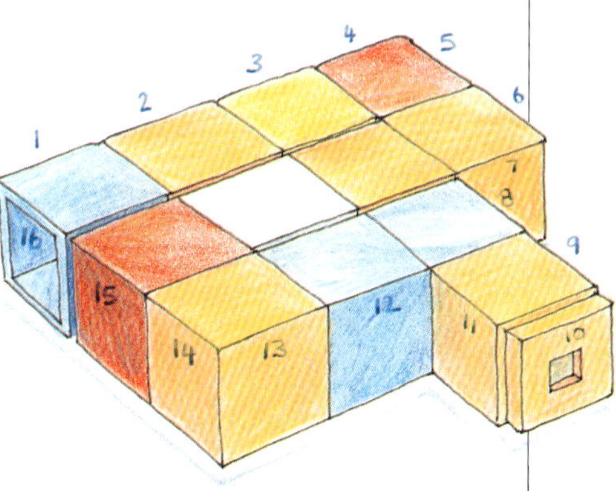

Lesson 3

Objective:

Students will practice finding the perimeter of a rectangle.

Materials:

UNIFIX CUBES

UNIFIX CUBE CORNER UNITS

Activities:

In this activity, students can work independently or in small groups.

Give each student, or group of students, a supply of *Unifix Cubes* and *Unifix Corner Units.*

Demonstrate how to create a rectangle using the corner units.

Ask the students to make various sized rectangles.

After the rectangles are made, ask the students to find the perimeter or distance around each rectangle. To find the perimeter, ask the students to count the outside edges of the cubes. A record is made for each rectangle.

For example, the students should write:

Perimeter = 14 units
or
P = 14 units

Then ask the students to make all the rectangles they can that have a perimeter of 24 units. Some of the rectangles will be eight cubes long and four cubes wide, six cubes long and six cubes wide, ten cubes long and two cubes wide, etc.

Ask, *"Were any of the rectangles squares?" "How are squares different from other rectangles?"*

Lesson 4

Objective:

Students will learn the formula for perimeter of a rectangle.

Materials:

UNIFIX CUBES

UNIFIX CUBE CORNER UNITS

SHEETS OF WHITE PAPER

Activities:

In this activity, students can work independently or in small groups.

Distribute a quantity of *Unifix Cubes, Unifix Cube Corner Units* and sheets of white paper to each student or group of students.

Ask the students to construct different sized rectangles with the cubes and then place the rectangles on a sheet of paper and trace the outer edge of the rectangle.

Students should then label each rectangle with the number of units long and the number of units wide.

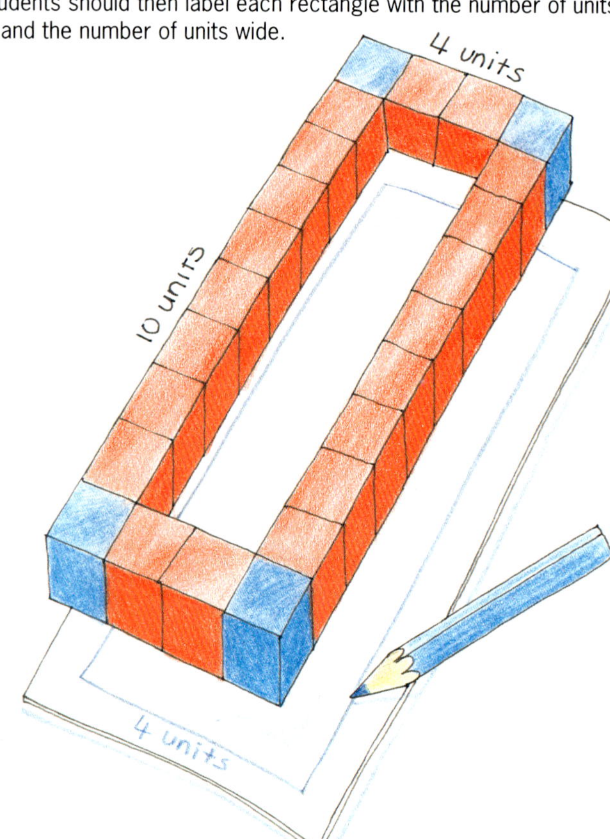

Discuss with the students the two ways they can find the perimeter of each rectangle.

Using the example of a rectangle that is ten units long and four units wide, the students could add 10 + 4 + 10 + 4. The other way is to multiply and add (2 x 10) + (2 x 4).

Write the formula **P = 2L + 2W** on the chalkboard and ask the students to use one of the formulas to find the perimeter of each rectangle they have traced.

P = 2L+2W

Lesson 5

Objective:

Students will gain an understanding that area is the number of square units that it takes to cover a surface and will learn the formula for the area of a rectangle.

Materials:

UNIFIX CUBES

UNIFIX ASSEMBLY GRIDS

UNIFIX RETAINING FRAMES

Activities:

Depending upon the size of the class, this activity may be conducted as an individual activity or students can work in small groups.

Give each student or group of students a quantity of *Unifix Cubes* and a variety of different sized *Unifix Assemby Grids.*

Tell the students that each cube has an area of one square unit.

Using the assembly grids, ask the students to construct a shape that would have an area of eight square units.

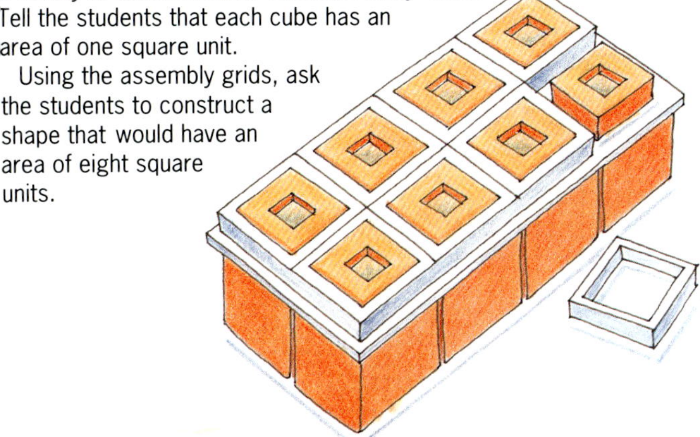

Next, the students using the grids and cubes should make rectangles that have an area of 12 square units.

Ask the students to trace around each shape and label them: *"Area = 12 square units."* Then ask the students to mark the length and width of the rectangles in number of units.

For example, 3 units and 4 units.

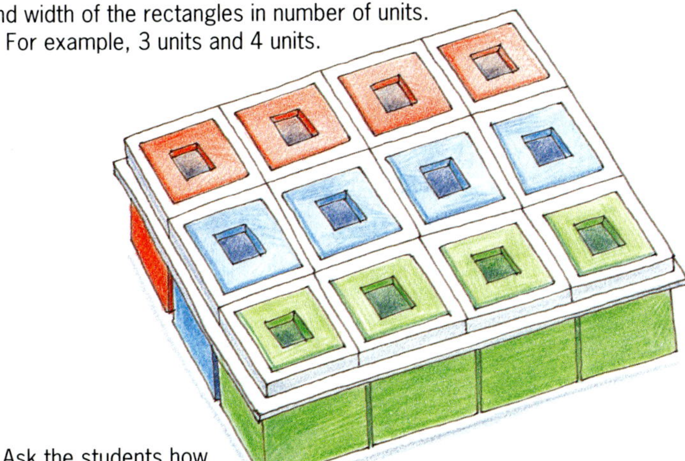

Ask the students how they can find the area of a rectangle if they only know the length and the width of the rectangle.

Once the students understand that **Area = Length x Width,** write the formula **A = LW** on the chalkboard.

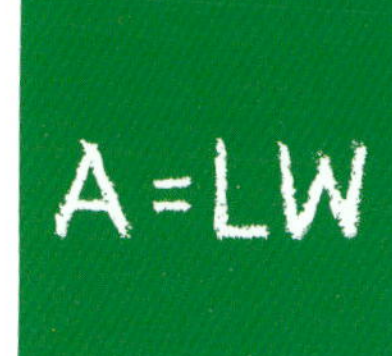

Continue working with the cubes and grids in creating different sized shapes and determining the area of each.

Lesson 6

Objective:

Students will gain an understanding of the concept of volume and of shapes.

Materials:

UNIFIX CUBES

UNIFIX ASSEMBLY GRIDS

UNIFIX RETAINING FRAMES

Activities:

Explain to the students that a Unifix Cube has the volume of one cubic unit.

Place students into groups and distribute a supply of *Unifix Cubes, Unifix Assembly Grids* and *Unifix Retaining Frames* to each group.

Ask the students to construct a model that has a volume of eight cubic units, using two rows of four cubes or four rows of two cubes with the Unifix Assembly Grid.

Continue the activity with students constructing two different models that have a volume of 12 cubic units.

Then ask the students to construct three different models that have 24 cubic units.

Looking at the three models of 24 cubic units, discuss with the students how the volume of the cube can be determined by multiplying.

15 Creating Your Own Unifix Support Materials

UNIFIX

Work or activity cards can be used to reinforce concepts that were previously developed with Unifix materials.

The cards can provide a pictorial model that becomes a bridge between concrete materials and abstract work with symbols.

The work cards can be used in many ways, such as for individual work for those students who need more reinforcement or they can be used in centers where students can take turns doing the work cards.

Following are a few ideas for making your own work cards. After making these cards, your own creativity will take over and more ideas will appear that will fill the needs of your particular students.

You will need the following materials to make and use reproducible activity cards:

- **Unifix Cubes**
- **Unifix Gummed Sheets**
- **Unifix Rod Stamps - Actual Cube Size**
- **Unifix Rod Stamps - Half Cube Size**
- **Unifix Number and Group Making Stamps**
- **Unifix Blank Underlay Cards**
- **Miniature Rubber Stamps**
- **Stamp pad**
- **White cardboard or heavy paper - 8 1/2" x 11"**
- **Unifix Wax Crayons**

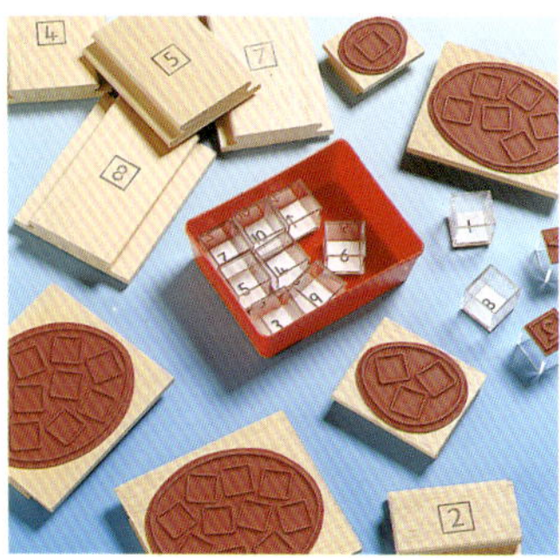
Number and Group Making Stamps

Cubes

Gummed Sheets

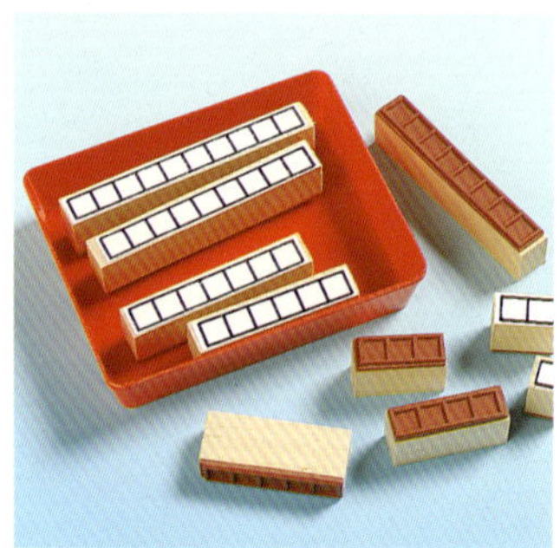
Rod Stamps

Blank Underlay Cards

Wax Crayons

Transition Boards and Unifix Cubes

Addition and Subtraction of Two- and Three-Digit Numbers.

Transition Boards or Mats can be prepared following the directions below for use by students in manipulating Unifix Cubes and towers to demonstrate trading of tens for ones.

The lessons in Chapter 8, "Addition of Two- and Three-Digit Numbers," and in Chapter 9, "Subtraction of Two- and Three-Digit Numbers can be implemented on Transition Boards or on the Blackline Masters provided for those lessons.

The Transition Mats can be prepared by a teacher aide or volunteer parents on 12" x 18" card stock or heavy construction paper and then laminated or covered with clear plastic contact paper.

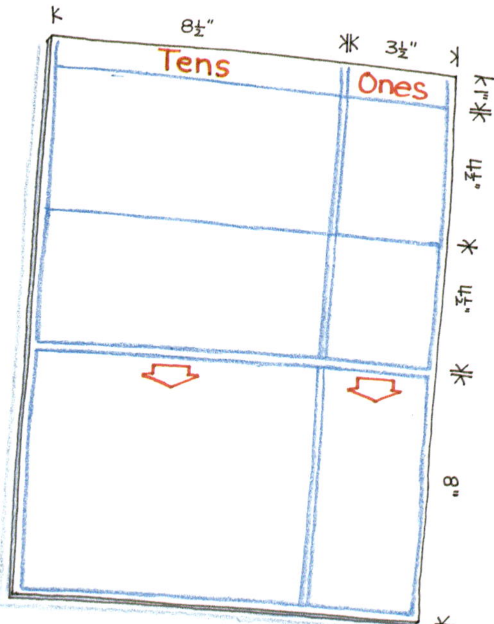

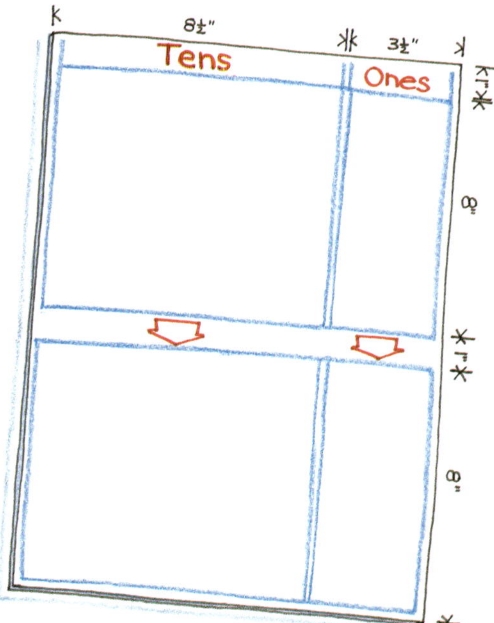

Early Ideas of Pattern

Make a set of ten reuseable pattern cards using Unifix Gummed Sheets. Students choose a card and make the pattern using Unifix Cubes and then copy the pattern on a sheet of paper using Unifix Wax Crayons.

Example:

Card 2 - red, red, yellow, red, red, yellow
Card 3 - red, yellow, blue, red, yellow, blue

The next set of cards can be made using the white Unifix Gummed Sheets with numbers.

Example:

Make a series of reuseable cards with the Unifix Gummed Sheets illustrating a variety of patterns. Students can select a card, cover the pattern with Unifix Cubes and then copy the pattern using Unifix Wax Crayons.

Example:

Complete The Pattern

Place a sheet of paper or cardboard in a horizontal position and create a variety of patterns or sequences using the Unifix Gummed Sheets. Students are to determine the pattern or sequence and use Unifix Cubes to create the next number. This workcard is not reproducible if the students are to duplicate both the number of cubes in the pattern or sequence as well as the color of the cubes.

Early Number Activities

Using Miniature Rubber Stamps, make student workcards in which students must match number symbols 1 to 5 with sets of objects by drawing lines from the number to the objects.

Using Unifix Group Making Stamps, make worksheets with sets of numbers 1 to 5 in which the students write the correct number next to each group.

Make Ten

This workcard will assist young students in understanding numbers to ten. Place the paper or cardboard horizontally and print: Stamp Out Number Ten.

Then, using the Unifix Rod Stamps, create a model of ten at the top of the page. Tell the students that they are to make as many combinations of numbers to make ten as they can using the Rod Stamps. Ater the students use the stamps, they can color in the stamps or cover them with the Gummed Sheets.

Additional workcards can be made for the numbers 8, 9, 11 and 12.

Number Between

On this workcard, use the Unifix Rod Stamps Half Cube Size to create pairs of rods in which the "Number Between" is missing. This activity will give students practice in the order of numbers. Tell the students that they are to use the correct Rod Stamp to fill in the missing number.

Number Before

Using the Unifix Rod Stamps Half Cube Size, make a workcard to illustrate "Number Before." Stamp pairs of rods with a line in front of each pair indicating that students are to find and stamp the number that comes before the pair.

A similar workcard can be made for "Number After."

Ways to Make Eight

Using Unifix Gummed Sheets, make a workcard that shows the different ways of making eight. Use two different colors for the two parts of the number. Students are to write the number fact following each set.

The same workcard can be made for the numbers six, seven, nine and ten.

Find the Sum

Use the Unifix Rod Stamps Half Cube Size to create a workcard that requires the students to find the sum of two rods and to stamp the correct rod after the = sign. Students should then write the entire addition problem, e.g., 3 + 4 = 7.

Workcards can be made for all the number facts up to the sum of 18.

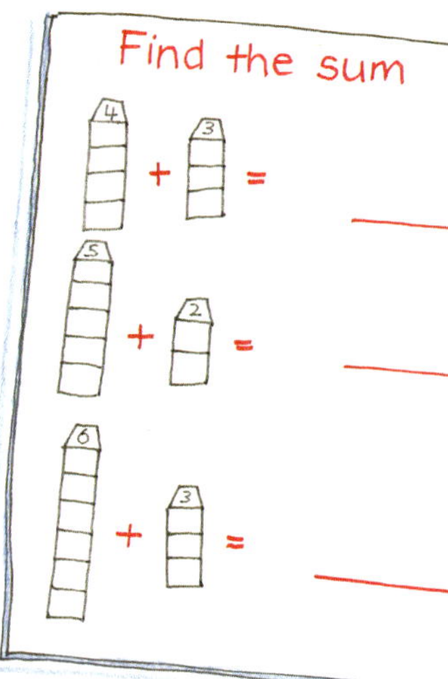

Difference Between

Prepare a subtraction workcard with the Unifix Rod Stamps Half Cube Size in which the rods are positioned to illustrate take away. Ask the students to find the correct Rod Stamp to complete the equation and stamp it after the = sign and then write the number fact.

Additional workcards can be made to show the subtraction facts up to ten.

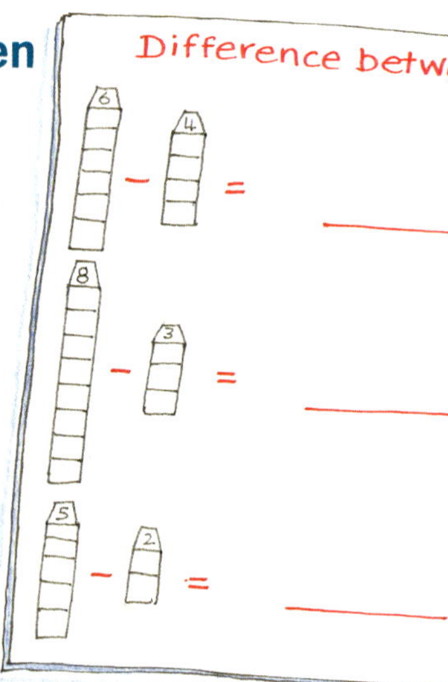

Take Away

Using Didax Miniature Rubber Stamps, create worksheets illustrating the concept of take away.

Then make a series of reuseable cards illustrating take away using Unifix Gummed Sheets.

Students have particular difficulty with the numbers 10 to 20 and need repeated practice at building up, naming and writing these numbers. The following activities offer suggestions for creating your own workcards and worksheets utilizing Unifix materials.

Tens and Ones

To reinforce the concept of two-digit numbers, make a worksheet with the Unifix Rod Stamps Half Cube Size of towers of ten and ones. Place the tens and ones in random order on the sheet and print under each set, "How many?"

Several worksheets can be made with various numbers of tens and ones and a card with all tens can also be made.

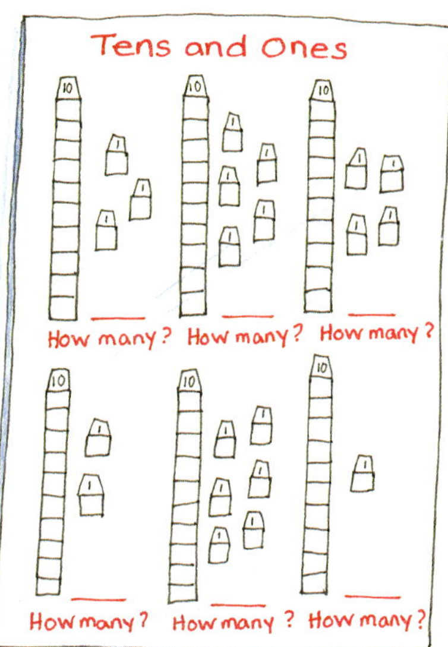

The Teens

Using the Unifix Rod Stamps Half Cube Size, prepare a worksheet for each student as follows:

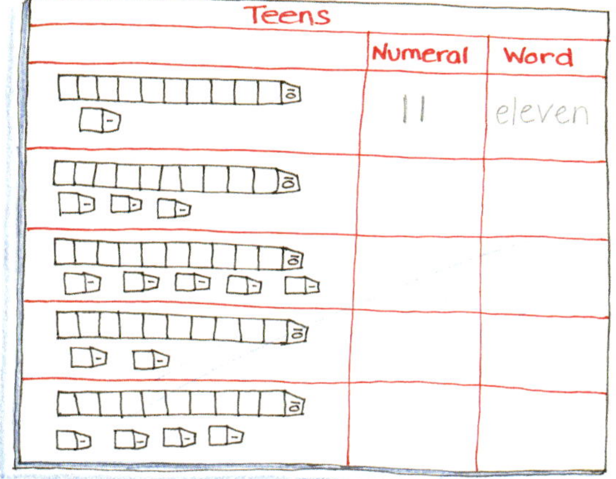

Early Multiplication

In the early stages of learning multiplication, it is helpful to make a set of cards which students can work on in pairs. These cards give practice in the concepts of multiple addition and grouping. Using the Unifix Number and Group Making Stamps, make a series of workcards similar to the following.

Times Tables Worksheets

Make a series of worksheets using Miniature Stamps which will demonstrate the times tables. For the two times tables use the bird stamp; nest with three eggs for three tables; horse with four legs; hand with five fingers; the hexagon for six; and two hands for ten.

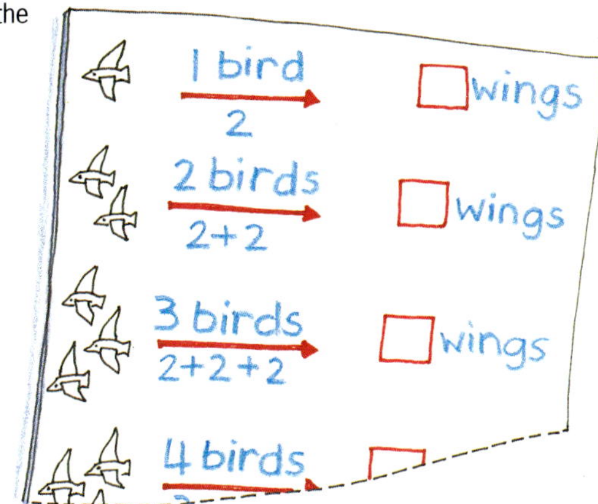

Linking Multiplication and Division

Using the Unifix Number and Group Making Stamps, prepare workcards similar to the example below to help students see the relationship between multiplication and division. Other numbers that can be used are 18, 21, 24 and 36.

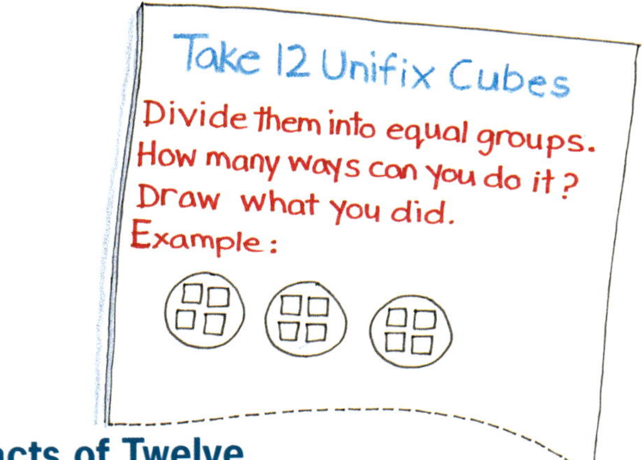

Facts of Twelve

By using the Unifix Gummed Sheets on a horizontal piece of paper, you can create a workcard that will help students learn the multiplication facts. Below is an illustration of how to present the Multiplication Facts of 12. Additional workcards can be made for other products.

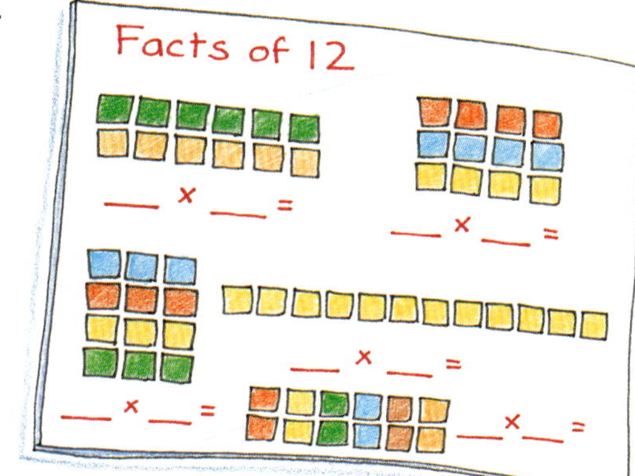

16 Teacher Resources - Aids in Learning Basic Number Facts

Additional Activities Using Blackline Master #1

Meaning of Addition

The following activities further develop the "Part + Part = the Whole" format of teaching the meaning of addition to young students. Unifix Cubes, Unifix Number and Group Making Stamps and numeral cards are used with the Addition and Subtraction Mat.

Example:
5 + 3 = ?
Students place Unifix Cubes in the upper parts of the mat and then move the cubes down to the "whole" portion of the mat and count the cubes to arrive at a sum.

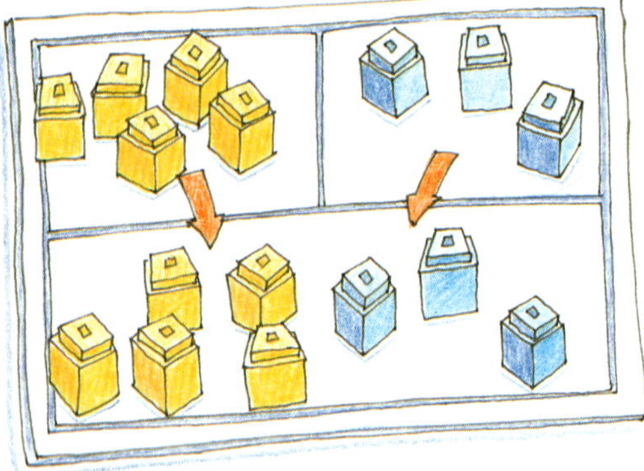

Example:
2 + 4 = ?
Next, using the Unifix Number and Group Making Stamps, ask the students to make picture cards of different groups of cubes.

Students should place the card with the groups of two cubes in the left part of the mat and actual Unifix Cubes in the other part.

They should then move the card and cubes down to the "whole" region and count the representation of the cubes and the cubes to find the sum.

Example:
4 + 6 = ?
The next step in the development of the understanding of the "Part + Part = Whole," would be to ask the students to use the cards they have created with the Unifix Number and Group Making Stamps on the blackline master.

Students move the cards down to the "whole" portion and count the sum.

Example:
4 + 5 = ?
In the next stage students are asked to use a numeral card as one "part" and actual Unifix Cubes as the other "part" to be moved down to the "whole" portion and added.

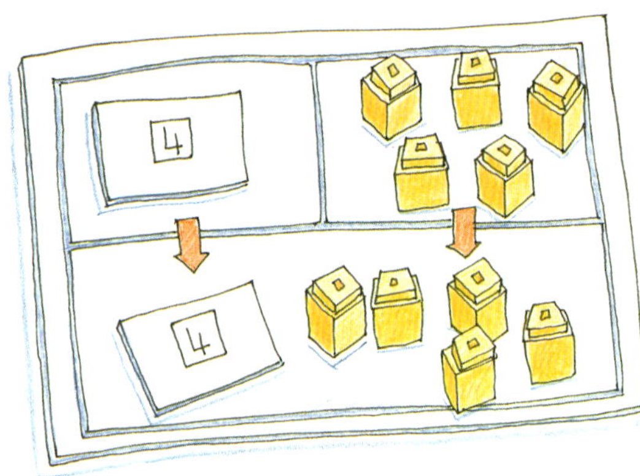

In the next stage, ask the students to use numeral cards and the picture cards created with the Group Making Stamps to illustrate the problem 3 + 4 = ?

The final stage is to ask the students to solve the problem 5 + 3 = ? using the numeral cards only.

A Logical Mathematical Model

The operations of addition and subtraction as well as the operations of multiplication and division are related and are inverse operations of each other. The diagram below illustrates that addition and subtraction are inverse operations, multiplication and division are inverse operations and that multiplication can be taught as repeated addition of like addends and division can be taught as repeated subtraction.

Connectors

Connectors link or bridge the concrete world of real things done manipulatively through hands-on activities with the abstract, mathematical symbolic world. They form the link between manipulative, representational or abstract forms.

Connectors are all of those schemes, materials, games and activities that teachers use to help students "see" into, and become competent in, the world of mathematics.

Connectors are concept helps that aid in giving meaning to abstractions. You may have students tell stories about numbers, do number dramas, have students draw number situations or build number related projects. Connectors are doing things in the student's world that give meaning to abstract ideas.

The following representations graphically illustrate how manipulatives connect the concrete and the abstract concepts of addition, subtraction, multiplication and division.

Addition

Level of Operation | **Countable Objects** | **Measurable Objects**

Manipulative:
Countable Objects: Join two sets, **usually unequal,** to form one set. Part, Part, Whole Concept.
Measurable Objects: Lay two lengths, **usually unequal,** end to end.

Representational:

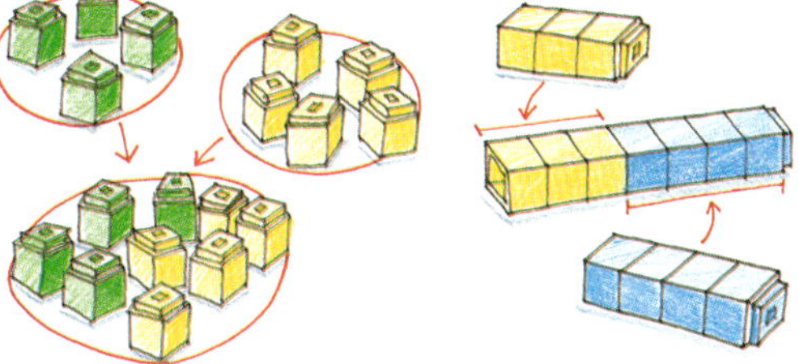

Abstract:
Words. "add", "plus", "and", "together", "more", "sum", "total", "whole", "addend", "join", "part", "answer", etc.
Symbols. 4 + 5 = 9 3 + 4 = 7

Subtraction

Level of Operation | **Countable Objects** | **Measurable Objects**

Manipulative:
Countable Objects: Comparison: match two sets in one - to - one correspondence.
Partition: Break set into two parts and remove one.
Measurable Objects: Comparison: match one set of end points.
Partition: Cut length into two parts and remove one.

Representational:
Comparison:

Comparison:

Partition:

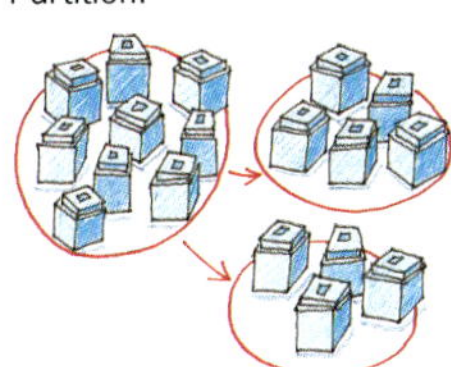

Partition:

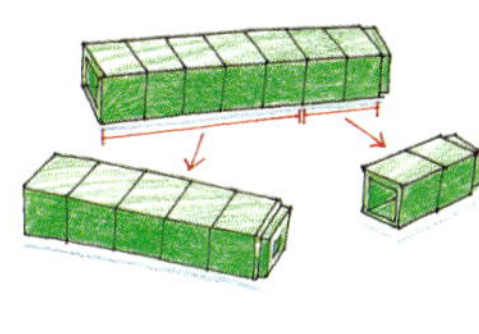

Abstract:
Language. "take away", "subtract", "minus", "less", "more than", "shorter than", "less than", "longer than", "answer", "more", "missing addend"
Symbols. 9 - 5 = 4

Multiplication

Level of Operation | **Countable Objects** | **Measurable Objects**

Manipulative:
Countable Objects: Join any number of sets, **all equal,** to form one set.
Measurable Objects: Place two lengths **perpendicular to each other** and complete the rectangle.

Representational:

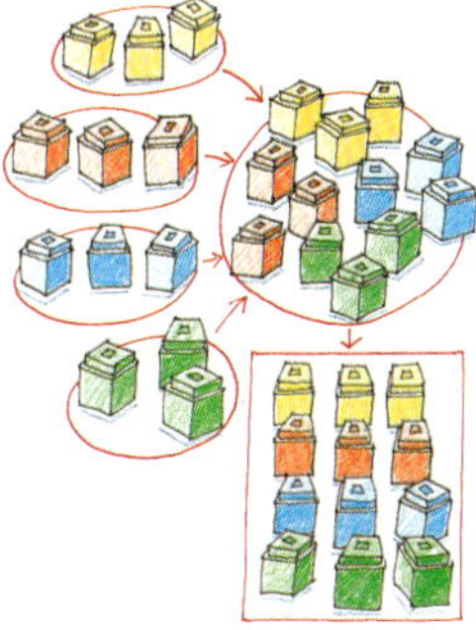

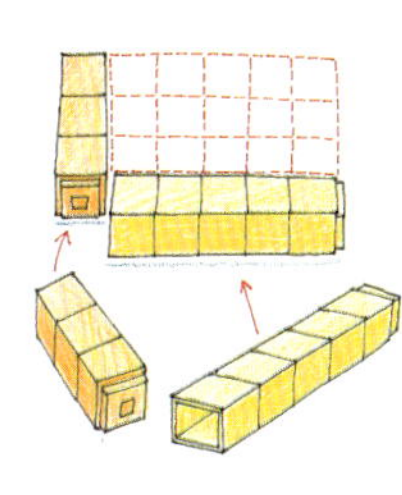

Abstract:
Words. "times", "multiplied by", "repeated addition", "factor", "product", "multiplier", "multiplicand"
Symbols. 4 X 3 = 12 3 X 5 = 15

Division

Level of Operation | **Countable Objects** | **Measurable Objects**

Manipulative:
Countable Objects: Partition into two or more equal parts
Measurable Objects: Partition into two or more equal parts

Representational:

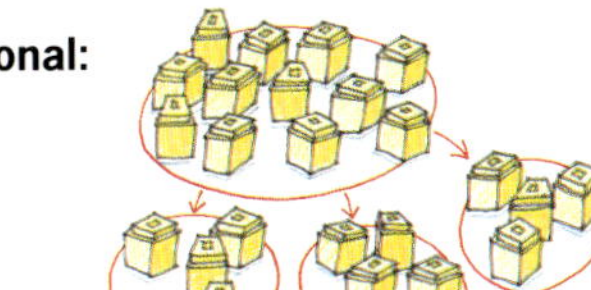

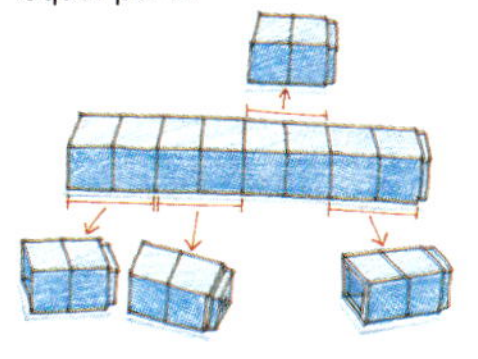

Abstract:
Words. "divide", "divisor", "dividend", "quotient", "give a fair share", "deal out fairly", "remainder"
Symbols. 12 ÷ 3 = 4 8 ÷ 2 = 4

$3\overline{)12}$ = 4 $2\overline{)8}$ = 4

Addition and Subtraction Mat

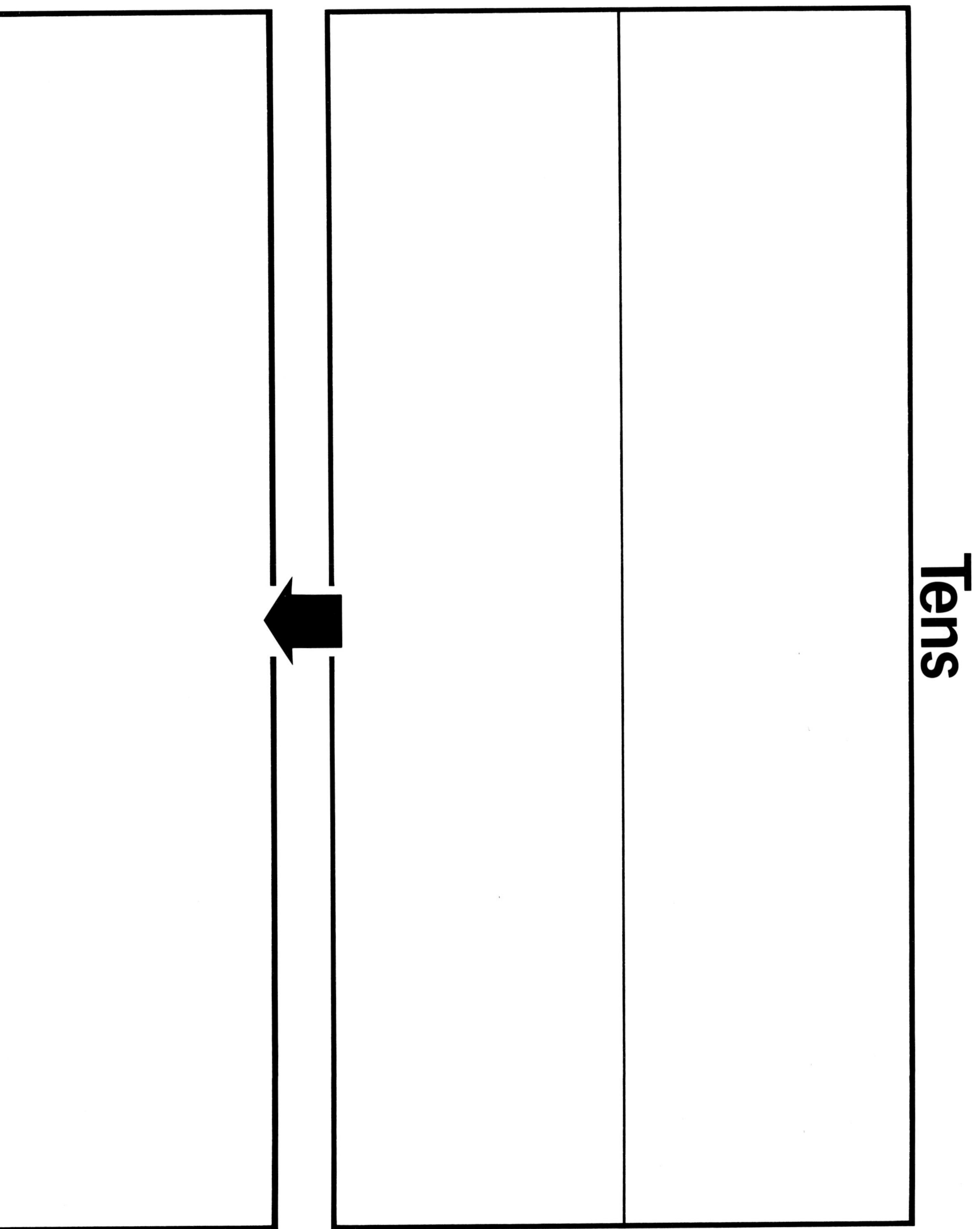
Tens

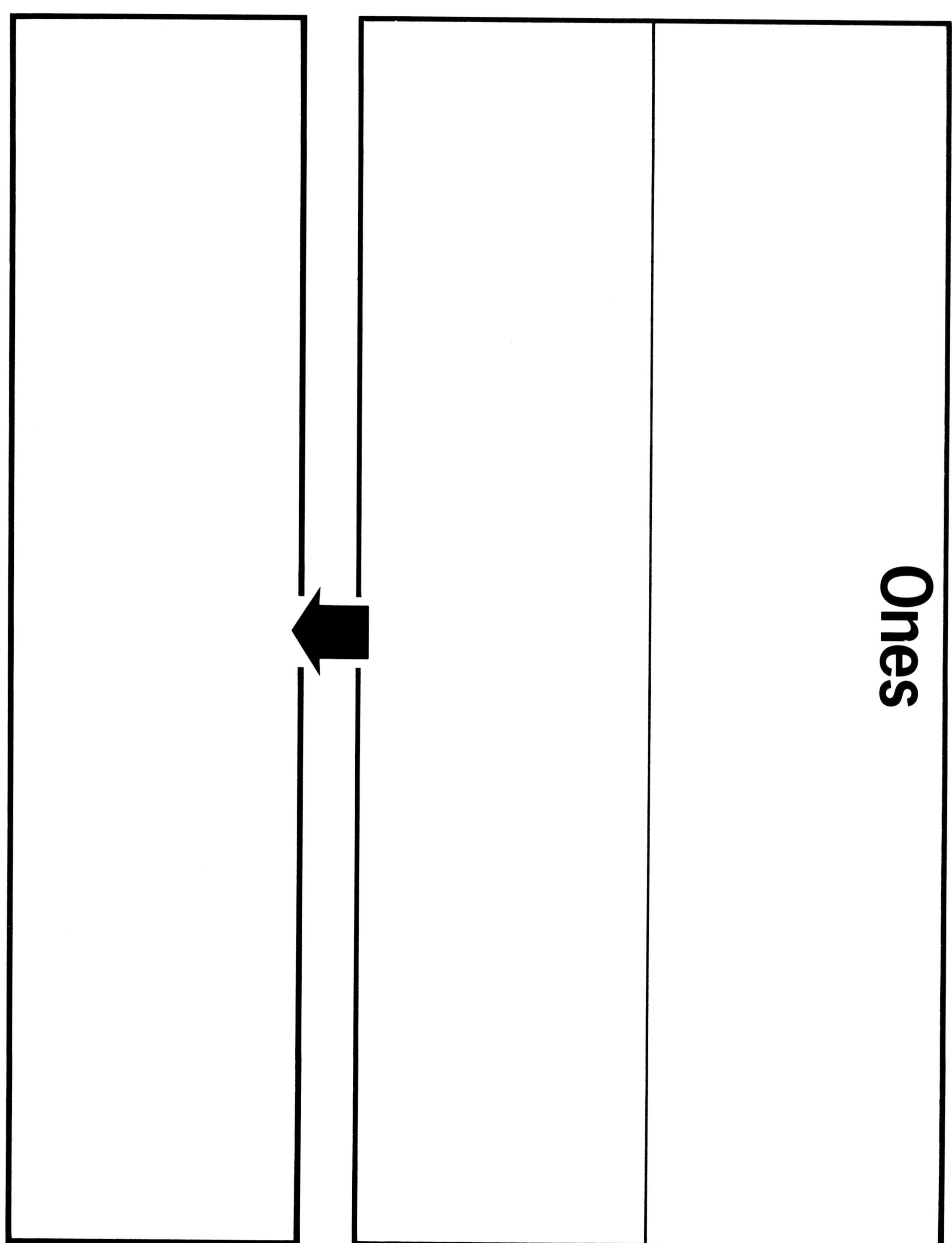
Ones

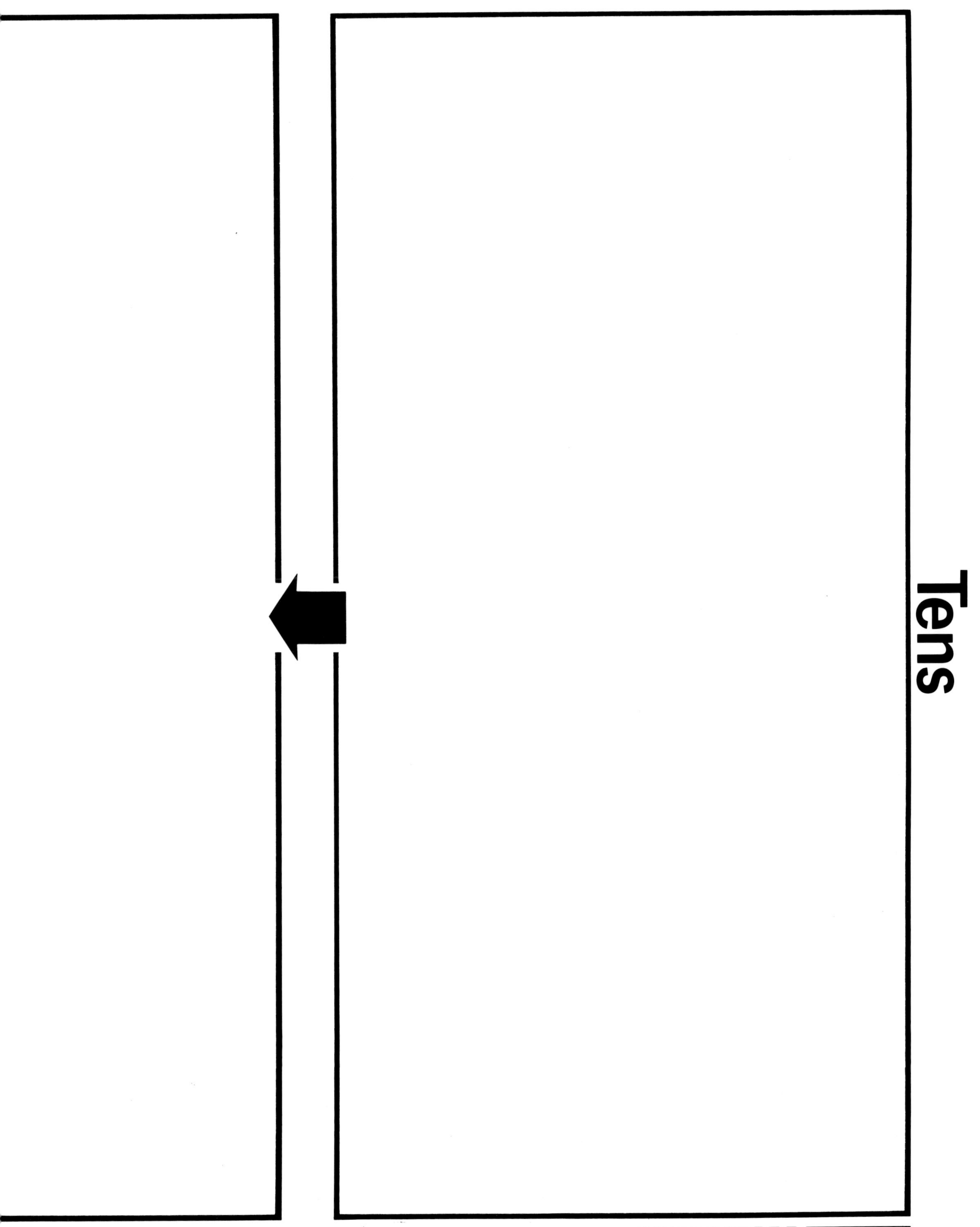
Tens

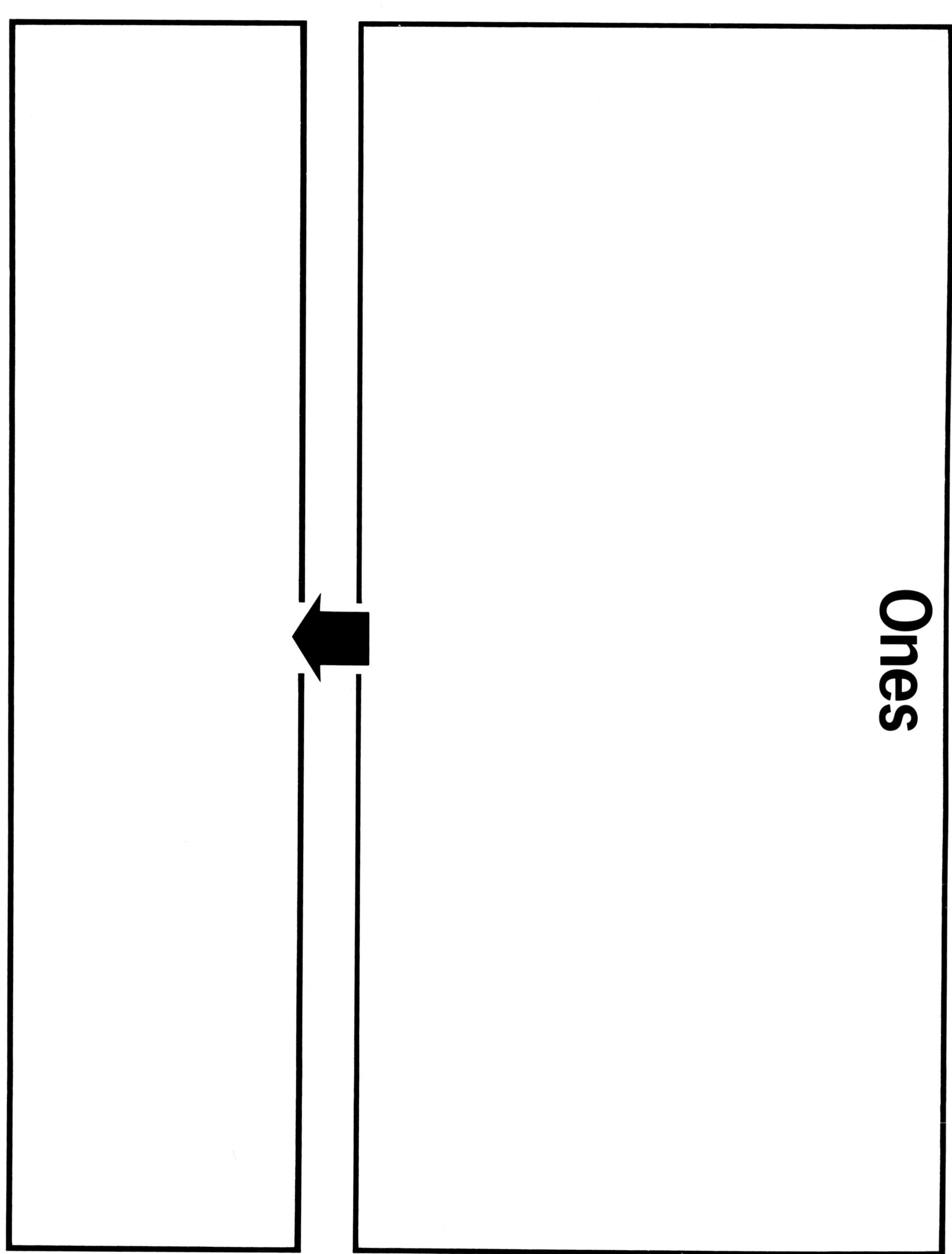
Ones

Hundreds	Tens	Ones

UNIFIX

Unifix Materials

Cubes

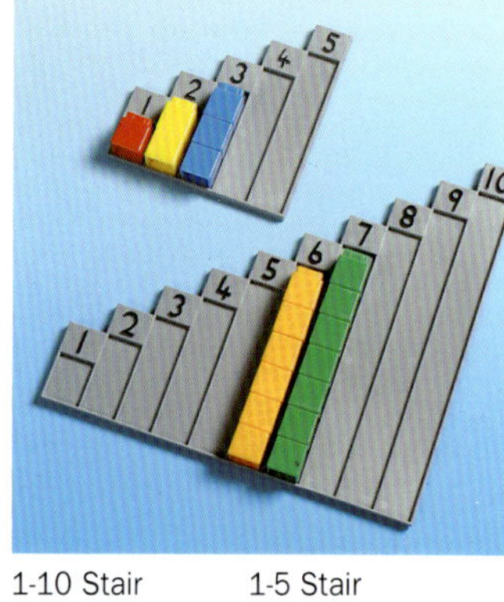

1-10 Stair 1-5 Stair

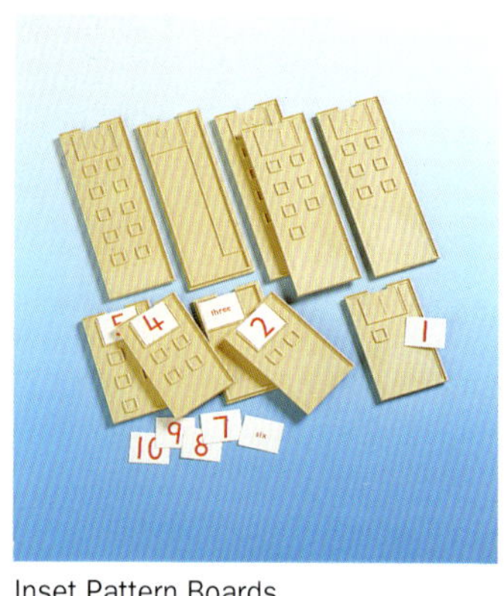

Inset Pattern Boards

10x10 Number Tray

Operational Grid and Tray

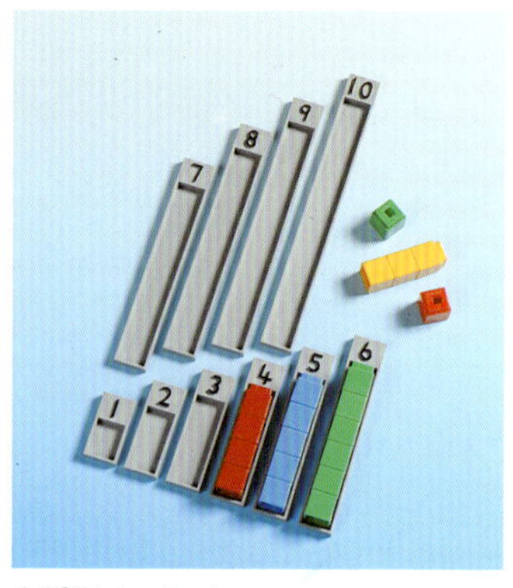

1-10 Value Boats

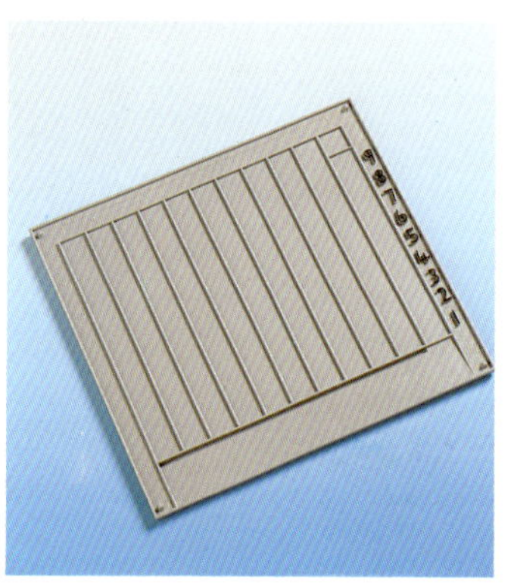

Dual Number Board

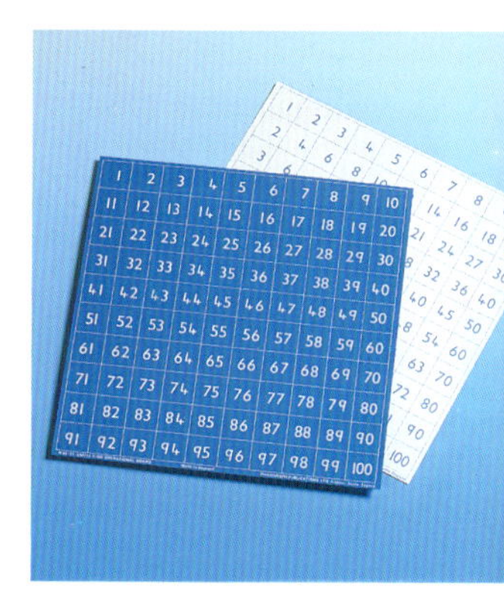

Underlay Number Cards

Pattern Building Underlay Cards

Cube Corner Units

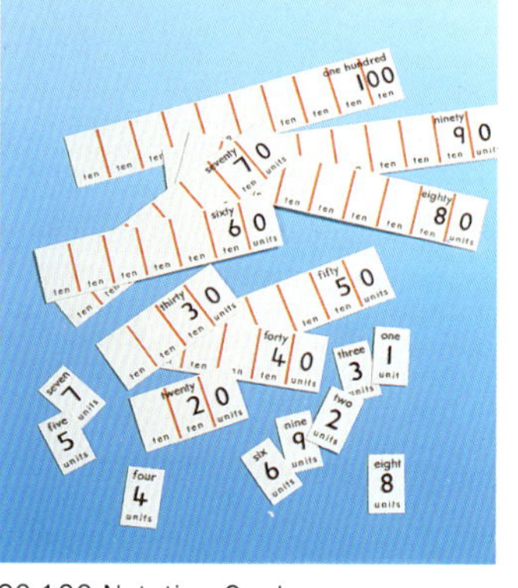

20-100 Notation Cards

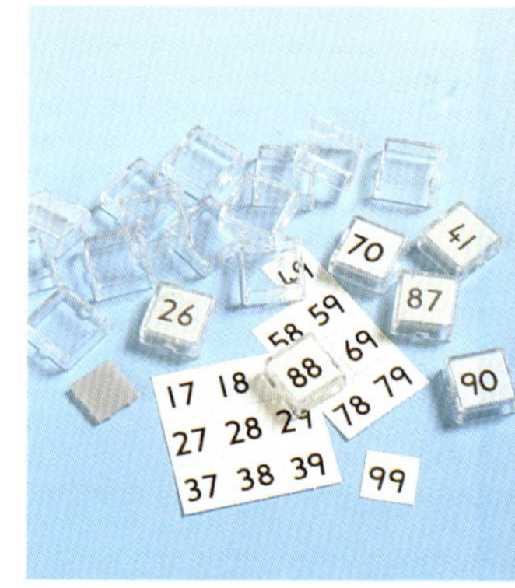

1-100 Number Tiles

Value Cards

My First Unifix Counting Book

Number Indicators

Window Markers

Unifix Materials

Building to 100 Board

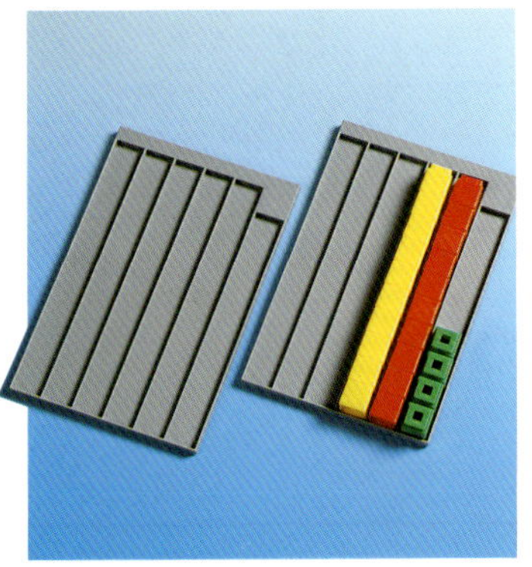
Five-Tens and Ones Tray

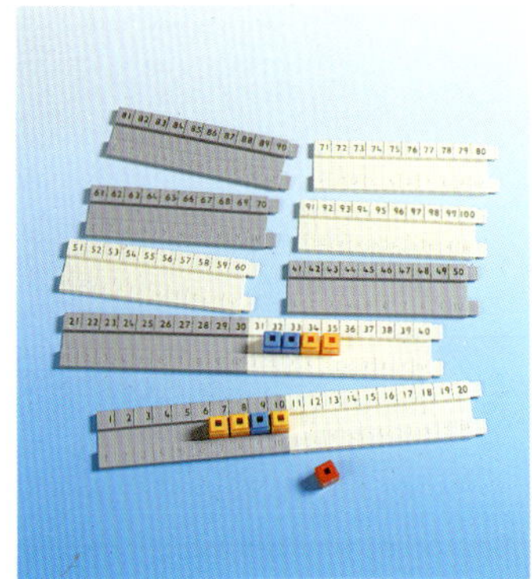
100 Track

Blank Underlay Cards

Cubes for the Overhead Projector

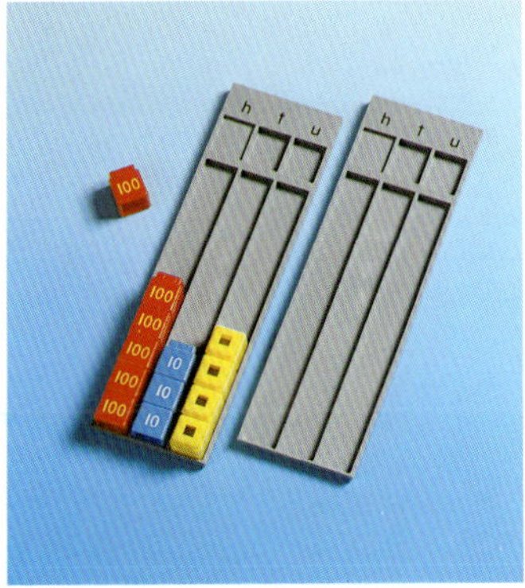
Hundreds, Tens and Ones Place Value Tray

Multiplication and Division Markers

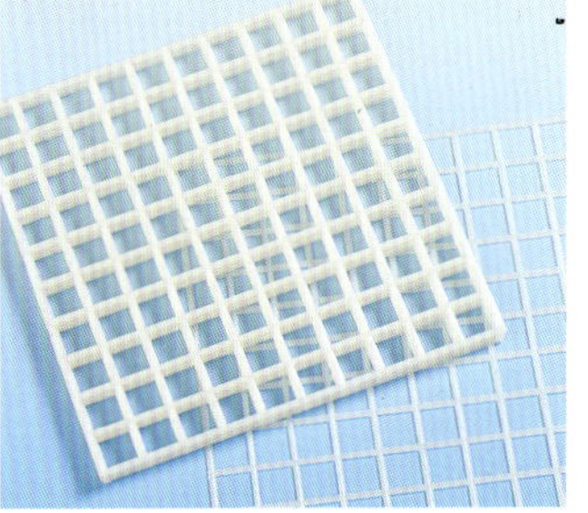
Retaining Frames & Assembly Grids

Wax Crayons

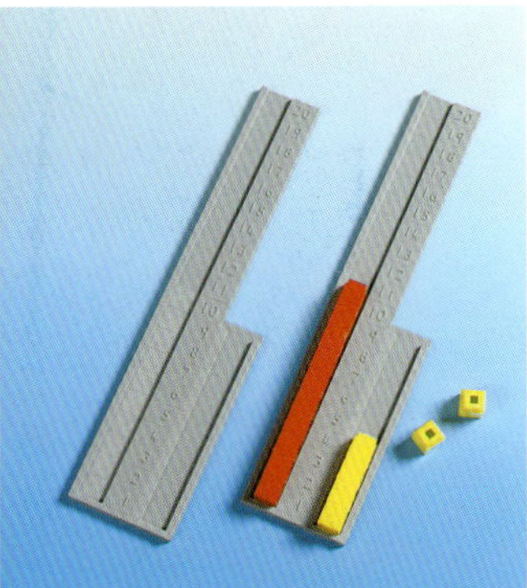
Counting Ladder

Tens and Hundreds Cubes

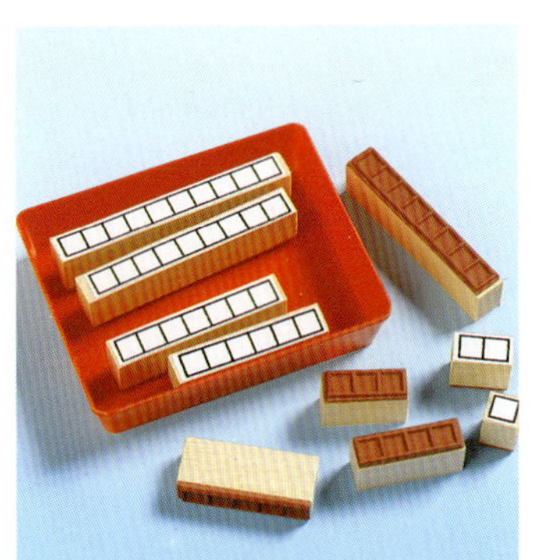
Rod Stamps

Gummed Sheets

One-Ten and Ones Tray

Number Cards

Blank Spinners

Number and Group Making Stamps

UNIFIX

UNIFIX

Index of Unifix Materials

Use this index to find the chapters and lessons in which the following materials are used.